OUR FOOD WILL BE FINE

Joris Relaes & Nele Jacobs

OUR FOOD

WILL BE FINE

ACADEMIA
PRESS

ILVO

Uitgeverij Academia Press
Coupure Rechts 88
9000 Gent
België

www.academiapress.be

Academia Press is a subsidiary of Lannoo Publishers.

ISBN 9789059964266
D/2025/45/621
NUR 740

Joris Relaes, Nele Jacobs
Our Food Will Be Fine
Gent, Academia Press, 2025, 168 p.

Cover: Studio Lannoo
Translation: Miriam Levenson
Layout: puurprint
Typesetting: Keppie & Keppie

CONTENTS

PREFACE

Our Food Will Be Fine is a book with a local character and international importance. At first glance it presents the research insights of a medium-sized Belgian government institute, Flanders Research Institute for Agriculture, Fisheries and Food (ILVO). Yet these insights flowing out of a tiny region of Belgium affect all of us, wherever we live and whatever we eat.

This book focuses primarily on Flanders, the densely populated, Dutch-speaking northern region of Belgium. Belgium is small but complicated: it has three official languages (Dutch, French and German) and three main regions: Flanders in the north, with intensive agriculture and strong food exports, Wallonia in the south, with extensive and often organic agriculture, and the region including Brussels, a cosmopolitan city with many farms located surprisingly close by.

ILVO is one of many research institutes working to make sure that our food will be fine. We invite you to join us as we contemplate food production in Flanders. Which elements of the Flemish situation resonate with your local circumstances? How do your insights about your own region, and the wider trends you observe, complement our point of view?

Like tasting a foreign dish for the first time, we wish you the pleasure of discovery as you savor the ideas in this book.

INTRODUCTION

Our food is just fine. You fill your plate at least three times a day, not with buttermilk, rye bread and potatoes like your ancestors did, but with a wide variety of healthy and even exotic foods. We rarely stop to think about it, but this is an extraordinary achievement made possible by a largely invisible food chain.

With more people now living in cities than in rural areas worldwide, the lack of urban food shortages is nothing short of a miracle. Even during the Covid pandemic, when we were all on lockdown, we sometimes ran out of toilet paper, but we never ran out of fruit and vegetables. The empty plate we serve you on the cover of this book symbolizes that daily *tour de force* while also standing for the freedom of choice we enjoy every day. Whether you're a dedicated carnivore, flexitarian, vegetarian, vegan, or pescatarian, and whether you eat seasonally, organically, or have food allergies, our well-oiled, flexible supply chain brings food from the farmer and fisherman all the way to your plate. With more than enough choice for everyone.

So if our food is just fine, dear reader, why read this book? Because agriculture, fisheries, and the food industry are at a crossroads—several, in fact. We can certainly expect a wave of challenges. But just around the corner are also opportunities, new insights, and groundbreaking technologies. Fact: climate change is forcing us to adapt quickly. Fact: artificial intelligence is creating new technical possibilities at warp speed. Fact: despite many pressures, farmers, fishermen and food companies continue to innovate, invest and do business. And another fact: you also influence the system by making certain choices when filling your plate.

Our food will be fine. We debated long and hard whether that sentence should end with a question mark. In certain ways, the current trend is more negative than positive. Every day, fertile farmland and open space vanish behind garden gates and under new bridges. The consequences of global warming are already weighing heavily on farmers, and global biodiversity continues to decline. Pessimists may feel vindicated by this, but they overlook the solutions currently being developed and the small changes that set larger transformations in motion. "Necessity is the mother of all invention," said agricultural economist Ester Boserup, as quoted in the final chapter.

Our food will be fine. This book is a forum for people at the forefront of change: Belgian scientists from Flanders Research Institute for Agriculture, Fisheries and Food (ILVO) and business leaders working on practical solutions. What can we expect from fermentation technology and regenerative farming? What is the potential role of data and robotics? What about measures focusing on water and soil? And what will happen with the new generation of plant breeding techniques and alternative protein sources?

Don't expect a tidy recipe book for the future, with answers to every question you might have. Expect a book that will make you think. This book is our invitation to you: let us test and refine some different recipes together.

Let's get to it! Wishing you a delicious read.

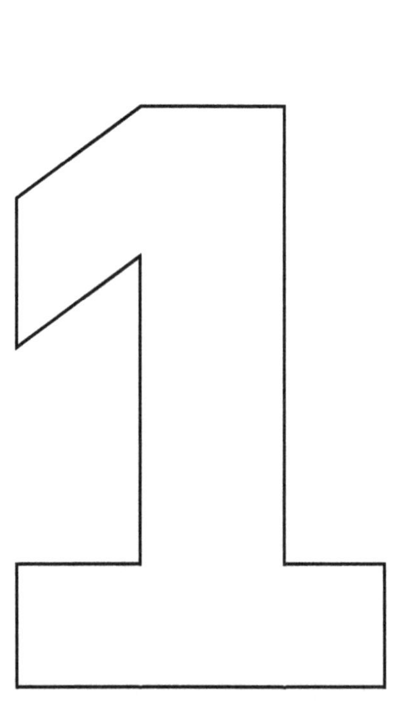

CHAPTER 1

Microbes: the Drivers of a New Agricultural Revolution?

The year is 2019. Covid-19 has not yet shaken confidence in globalization, and Ukraine—a major global supplier of grain and fertilizer—is not yet in the grip of war. Climate change is high on the political agenda: tens of thousands of young people are skipping school on Thursdays to protest in Stockholm and Brussels. In France, the "yellow vest" protests sow chaos and destruction, while in the Netherlands, public frustration is erupting into farmers' protests against the nitrogen policy.

In this tense but otherwise relatively stable context, scientists turn their attention to an American report titled *Rethinking Food and Agriculture 2020–2030*, subtitled *The Second Domestication of Plants and Animals, the Disruption of the Cow, and the Collapse of Industrial Livestock Farming*.

The end of intensive livestock farming, as it has been known for the past half-century, gets predicted with clockwork regularity in debates and on the opinion pages. But this report was different. What shocked the readers was not the technology described, but the speed at which that technology could become affordable. By 2030, it claimed, producing milk proteins in steel vats using microbes would be five times cheaper than producing them from cows. These low-cost microbial proteins would flood the American food chain as an alternative to cow's milk, triggering ripple effects far beyond the United States.

It is now almost 2030, and the predicted economic tipping point has not yet arrived. Even so, the report set a great deal in motion in Europe as well as in the US. The question is no longer *if* the turning point will come but *when*, and how the affected sectors will adapt.

What Are Proteins?

Proteins are essential building blocks of the human body. On average, however, most Westerners consume more protein than they need. We get proteins from animal products, but also from bread, pasta, potatoes, legumes and more. For our health, the source of protein matters as much as the total amount we eat. According to the Flemish Institute for Healthy Living, our intake should be roughly half animal-based and half plant-based.

Yet the "EI-MEET" protein monitor of the Flemish Government's Department of the Environment shows that, in 2023, the average Flemish diet still derived 58.7% of protein from animal sources and 41.3% from plant or mixed sources. Strikingly little—just 6.9%—came from so-called "high-quality" plant proteins such as legumes, nuts, Quorn and other meat or dairy alternatives. We dive deeper into alternative proteins in a later chapter.

Fermentation

In Ghent, a company called Those Vegan Cowboys is already doing what the American report predicted: they make soft cheeses in a lab without a trace of animal protein. These cheeses are indistinguishable from classic Camembert or mozzarella in taste, texture, and mouthfeel. Just a few kilometers down the road, Biotalys is leading a revolution in crop protection. Their innovation: camelid antibodies for plants to fight fungal infections, a potentially powerful alternative to chemical fungicides.

What do these neighbors have in common? They use no milk, no camels, and no synthetic chemical building blocks. Instead, they grow microbes in small bioreactors. These microorganisms are yeasts, bacteria, fungi, or microalgae—invisible to the naked eye, but essential to all life and all processes on Earth. The unique talent they all share? The ability to convert one molecule into another, which changes the properties of the environment where they live. Add lactic acid bacteria and salt to meat, and it ferments into salami. Add starter culture and rennet to milk, and it curdles into cheese. Add yeast to grape juice, and it loses its sweetness and transforms into wine.

The name of this transformational process? Fermentation. We have used it for centuries to preserve food. You can imagine that scientists have always been fascinated by fermentation. In their attempts to understand it, they have slowly uncovered a vast potential. Now we know that microbes and fermentation are behind the miracle that happens in a cow's rumen, the first of its four stomachs. As the cow eats grass and chews its cud, microbes in the rumen ferment the grass, creating milk proteins our bodies can absorb. Behold the magic of fermentation!

The logical next question is whether this fermentation process can happen outside of the cow. Can microbes in a lab directly produce nutrients that humans need? The answer is yes, and that is the game changer. With the precision fermentation technology used by Biotalys and Those Vegan Cowboys, it is now possible to produce virtually any protein, as well as most fats, flavorings, vitamins and pigments. Just put the right microbe in a controlled environment, then add water, energy and a suitable nutrient medium. Of course, the reality is more complex—we'll get into the details shortly—but the first products of this revolution are already on the market, and they're here to stay. One example is EverSweet from Avansya, a joint venture between food giant Cargill and ingredient manufacturer DSM-Firmenich. EverSweet is as sweet as stevia but without the bitter aftertaste. Available on the B2B market in Europe since 2024, it is about to transform the candy and beverage aisles.

Cultured Meat?

The new wave of fermentation technology may remind you of meat grown in a lab, but there is a key difference. Cultured (cell-based) meat starts from existing animal cells, which then multiply until they form muscle tissue. In contrast, microbial production uses no animal cells at all. Here, microbes are programmed to produce cells that take over the function of animal cells—taste, texture, color, nutrients—in the finished product.

Investors Are Enthusiastic

According to a study commissioned by Avansya, the carbon footprint of its EverSweet sweetener is 81% smaller than that of sugar. Its production would use 97% less water and 96% less land. This was only one study, but it illustrates why both scientists and investors are excited about this technology. It takes little land, few raw materials and far less time and water than traditional livestock farming or manufacturing, while emitting fewer greenhouse gases. Some of the biggest problems we face in food production—shortages of raw materials, nitrogen emissions, climate change and soil depletion—could be addressed, at least in part, by fermentation technology.

Full life cycle analyses (LCAs) are still needed for each molecule produced, but the production process shows clear potential for being highly efficient, cost-effective and scalable. Combine that with Europe's mandatory sustainability reporting (CSRD) for large companies and lenders, and you can see why investors are interested. Biotalys raised nearly €53 million in 2021 to develop its antibodies. Paleo, a Belgian start-up that uses yeast to produce the protein that gives meat its signature color and flavor, secured €12 million in 2023. The Biotope incubator of the Ghent-based biotechnology institute VIB has recently seen a steady stream of entrepreneurs seeking funding for their microbial innovations.

Venture capitalists are not the only ones making these moves; established names from traditional industries are stepping in as well. We already mentioned Cargill, which was built on global grain trading, and DSM-Firmenich, known for plastics, fragrances and flavorings. But Those Vegan Cowboys say that even conventional dairy and food companies are showing interest. "Food companies are facing increased uncertainty in supplies of raw materials and they are exploring ways to address that," says CEO Hille van der Kaa. "Others want to expand their range with plant-based alternatives." The chocolate sector is showing similar curiosity, with rising interest in cell-based cocoa as an alternative to costly cocoa beans.

Precision or Biomass Fermentation

Back to our cow. The microbes in its rumen produce a remarkably complex milk protein. In contrast, the microbes in your sourdough starter do nothing more than multiply on their own. Differences in complexity and end product can also be found in modern fermentation technology.

In single-cell or biomass fermentation, microbes are grown under controlled conditions, and the entire resulting paste or slurry is used as a protein-rich raw material. Quorn has been made this way for years. The process used by Those Vegan Cowboys and Paleo—precision fermentation—is far more complex. Here, microbes are programmed to produce a specific protein, fat or antibody with specific properties using gene-transfer technology. Lots of time, money and research are being invested into this step of the process. Thousands of microorganisms and millions of their strains first need screening, and then targeted genetic modifications are carried out to ensure they yield exactly the desired molecule. At the end of the process, only that specific molecule is harvested. The rest of the brew, including the biomass of the programmed microbe, is left over as a potentially usable by-product.

More Than Protein

Precision fermentation is far more complex and costly than biomass fermentation, but it attracts the most investment because the target molecules are far more rare, and far more valuable, than a protein-rich paste. Pioneering companies are racing each other to discover the one molecule so rare that the market will pay a premium for it. The initial hype centered on proteins, but investors now expect to see the first commercial breakthroughs elsewhere, in more scarce components required for food or feed ingredients, or for chemicals, pharmaceuticals and cosmetics. These could be substances we currently import from other regions, like amino acids or palm oil; compounds we produce synthetically, such as vitamins; plant-derived extracts like stevia; or substances that plants can hardly produce, such as certain pigments or camelid antibodies.

"Today, it costs three times as much to produce one ton of protein from fermented potato peels as it does to import one ton of soy from South America. Even swapping just 5% of soybean oil with an alternative—whether plant-based or microbial—would already triple production costs. Imported soy is simply too cheap to replace on cost grounds alone," says Annatachja De Grande, ILVO researcher and feed specialist. She recognizes the lack of appetite for microbial protein in the feed sector, but amino acids and fats from microbial sources are another story. "Feed producers are heavily dependent on imports from the U.S., South America and China for amino acids. The sudden price spikes and shortages of recent years have shown just how vulnerable those imports make livestock farming. We are self-sufficient in fats for feed, but microbial fats could improve the quality of animal products. It's possible to tailor the fatty acid profile in meat or eggs to meet the needs of specific consumer groups. Retailers and pet food producers are especially interested—margins in dog and cat food are higher than in livestock feed. Who knows? The first breakthrough might just happen there."

Flanders at the Forefront

Perhaps surprisingly, the tiny region of Flanders has become a global hotspot for fermentation technology research. Companies from across the world bring their microbes to the Flemish Bio Base Europe Pilot Plant for help in scaling up production. Knowledge centers Ghent University, KU Leuven and VIB lead the way in screening, designing and programming microbes to produce exactly what industry needs. Flanders also hosts the founders and members of the unique platform called The ProteInn Club, which concentrates high-level expertise in real-world applications ranging from food, feed and pet food to chemicals and cosmetics.

The ProteInn Club was founded in 2022 by ILVO, Bio Base Europe Pilot Plant, Ghent University and Capture, a research facility for the circular economy. The aim? To build a dynamic ecosystem for developing new value chains around fermentation technology in Flanders. Barely two years after its founding, the platform gathered some 200 professionals from across Europe to Merelbeke-Melle, a suburb of Ghent, to see lab- and pilot-scale innovations firsthand. The response was so strong that registrations had to close early due to the overwhelming demand for tours of the Bio Base Europe Pilot Plant and ILVO's pilot facilities. "Everyone is looking for the holy grail that will make their production more sustainable and circular," says relationship manager Stef Denayer. "Fermentation technology has that potential. What we're seeing today is just the tip of the iceberg."

The tip of that particular iceberg is quite impressive, however. Those Vegan Cowboys, Biotalys and Paleo are just a few of the Belgian start-ups that are quickly advancing in their development of microbial raw materials. The Scottish company Enough uses residual sugars supplied by Cargill to produce a fungal protein which is then used in Unilever's next-generation substitutes for meat and fish. Ghent-based Biolynx uses fungi to produce proteins for meat and cheese alternatives. They even use the same technology to make a UV filter for sunscreen. Avecom, a Ghent University spin-off, applies fermentation to convert food industry by-products into high-quality raw materials. In Jupille, Belgium, where Jupiler lager first flowed out of the vats, beer giant AB InBev is now brewing a range of protein-rich powders and sports drinks.

A new type of company is showing interest in The ProteInn Club, potentially transforming the entire fermentation sector: AI and robotics companies. The German firm VCG.AI, for example, develops models to automatically match technologies, raw materials and applications. Artificial intelligence can also accelerate screening of microbial strains and genetic modification. "The combination of AI—which can rapidly detect patterns in all kinds of data—with robots that automate lab work has the potential to cut weeks, months or even years from the extensive biotech development phase. That's where many fermentation pioneers are today," says Karen Verstraete, communications officer at The ProteInn Club.

Market Approval

Even with AI and robotics speeding things up, it may still take years before these innovations reach the market. In Europe, food ingredients from companies like Paleo and Those Vegan Cowboys fall under Novel Food legislation, which covers any ingredient not consumed in the EU before May 1997. Such ingredients can only be sold after European

Commission approval, based on a safety assessment by the European Food Safety Authority (EFSA). This process, which takes years to complete, is based on a thorough food safety assessment.

"This process is vital," explains Lieve Herman, an ILVO department head, researcher and expert scientist at EFSA. "Some microbes are known to produce toxic substances under certain conditions. The application must demonstrate an understanding of those risks and how they will be controlled. These safety assessments are so extensive because the risks vary with every strain, every application and every realistic dose that a person would consume."

Those Vegan Cowboys are currently seeking approval in both Europe and the United States, and expect to launch their vegan cheeses and vegan milk proteins within a few years. Biotalys' antibodies also face a thorough evaluation under legislation for new crop protection products. They expect a 10-year waiting period before their product could reach the market.

Economical and Circular

Microbes growing in steel bioreactors need remarkably few inputs to thrive: just some energy, water and a source of carbon and nitrogen. That source can be either glucose or, at least in theory, by-products such as potato processing waste, whey from cheese production, molasses from sugar factories or beer draff, a grain-based by-product. The only requirement is that they are available in large volumes and are not a critical link in an existing processing chain. In some cases, they need pre-treatment—heating, grinding, removing unwanted compounds—and not all microbes grow on all feedstocks. But it's still worth exploring the possibilities. "Food industry by-products often go into animal feed, which is already a good use for them, but in some cases, they can be upgraded through fermentation," says ILVO scientist Geert Van Royen. "If that can yield biomass and molecules that could replace fossil-based raw materials, for example, that would be a win-win for the companies and for society."

Food and Feed

As a founding member of The ProteInn Club, ILVO has invested heavily in expertise and pilot infrastructure. The first cornerstone was already in place: the Food Pilot has a unique combination of equipment and flexibility to produce virtually any food application on a pilot scale. Located on ILVO grounds in Merelbeke-Melle, Belgium, the Food Pilot is operated jointly by ILVO and Flanders' FOOD, the food industry cluster of the Flemish government. This collaborative arrangement ensures that all investments directly benefit the sector. Recent additions include a small fermenter for producing limited batches of microbial protein specifically for research trials and a new protein processing line to dry, purify and turn pastes and slurry into nuggets, vegan mayonnaise, snack bars and other products.

Specialized ILVO labs analyze these new proteins. What is their chemical composition? Are they easily digestible? What are their functional properties, such as foaming or gelling capacity? "We also study the food safety of products made with new proteins as well as their shelf-life. We have a sensory lab for objective assessment of mouthfeel, taste and aromas. That combination of infrastructure and expertise makes us unique in Flanders and far beyond," says Geert Van Royen. He is part of a team exploring opportunities for new ingredients and processes. They collaborate with The ProteInn Club partners as well as agronomists, livestock experts and feed specialists at other ILVO sites. "We work from field to fork, and we mean that quite literally," Van Royen says. "In the field, my ILVO colleagues grow new protein-rich crops like chickpeas and faba beans (also known as fava beans, broad beans and field beans). In the Post Harvest Pilot—a recent addition at ILVO—we dry, sort and clean protein-rich seeds and other plant parts to prepare them for further processing. In the Food Pilot, we test their potential for human nutrition; in the ILVO Feed Pilot—another recent investment—we explore their use in animal feed. We do this for all kinds of new proteins and fats, whether they come from microbes, plants, insects, algae or by-products."

What Do Proteins Do in Food?

Proteins are complex molecules that play a key role in food quality. They are responsible for foaming in creams and drinks, they form gels in puddings and jams, and they bind ingredients in sauces and dressings. Depending on the molecular structure of the protein and the material it interacts with, it can add body, texture and stability to the final product. The way a protein is produced or isolated also influences how it behaves in food.

Life Cycle Analysis

Every piece of equipment in ILVO's three pilot processing plants–the Post Harvest Pilot, Food Pilot and Feed Pilot–is fitted with water and energy meters. This allows ILVO to assess not just the functionality of new proteins and fats, but also their environmental footprint. Unlike traditional animal products such as meat or dairy—where most of the environmental impact lies in the field, the barn, or in the bioreactor itself—microbial proteins shift that impact to later stages, such as purification, drying or texturing. "Life cycle analysis allows us to track environmental impact across the entire chain. In that way, we avoid shifting problems to other parts of the process or other parts of the world," explains Veerle Van linden, an environmental sustainability expert at ILVO.

Life Cycle Assessment (LCA)

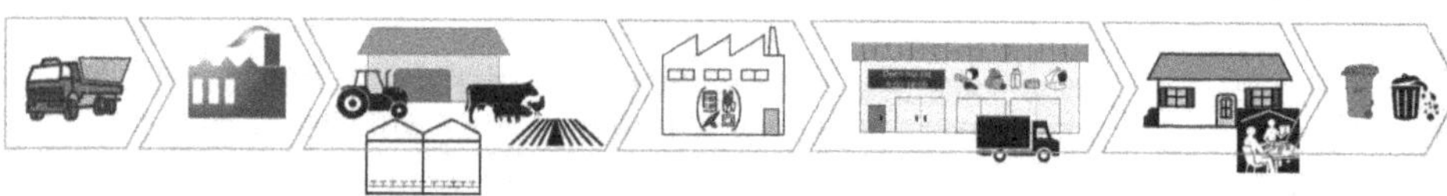

Figure 1: Life Cycle Assessment calculates the total environmental and climate impact of a product. It starts with raw material extraction and transport, and ends with consumer use and waste disposal. Per "link" in the chain, an LCA measures the amount of water, energy, land, metals, minerals and fossil resources used, as well as the emissions produced. This gives a good estimate of how the entire production process affects human health, the quality of our ecosystems and the depletion of non-renewable resources. © ILVO

This brings us back to the cow and the possible impact of fermentation technology on livestock farming. The 2019 U.S. report overlooked one important thing: cows do more than simply convert grass into milk and meat. Until recently, this was the main selling point of the dairy cow, also shared with goats and other ruminants. While ruminants may lose their monopoly on milk production, they still produce collagen, rennet, leather, hide glue, Botox, fat and manure. They help manage landscapes and keep our fields fertile, and they are the main reason for us to maintain valuable grasslands, which capture as much carbon from the air as forests and nature reserves do. Cows also process millions of tons of food industry by-products each year. At the opening of ILVO's Feed Pilot in June 2025, Katrien Deschepper, director of the Belgian mill Paniflower (part of La Lorraine Bakery Group), emphasized the importance of this valorization. "Every year, we sell 80,000 tons of by-products for processing into animal feed. That's an enormous amount and an important additional market for our company."

Growing one milk or meat protein in a steel vat cannot compare to the complex value chain of dairy production. The trumpeted "end of the cow" is not likely to happen. Still, livestock farming is vulnerable to market forces. Animal products and the raw materials required for animal farming are relatively cheap but highly volatile. Prices surge at the slightest shortage and collapse with oversupply. If it is indeed true that a surplus of cow-related products (milk proteins, collagen, fat, etc.) would occur due to competition from microbial proteins, this could indeed threaten the profitability of dairy farming.

"Protein is the ingredient at the economic cornerstone of dairy production. But fresh cow's milk contains only 3.3% protein. The rest is water (87.7%), lactose (4.9%), fat (3.4%), plus vitamins and minerals (0.7%)," wrote the American authors in the report quoted at the beginning of this chapter. That raises a few eyebrows, because large-scale microbial production of casein, the key milk protein, is looming on the horizon. Those Vegan Cowboys are already doing it at lab scale. The eventual market balance is hard to predict, but the effect of microbial protein will most certainly ripple through the pre- and post-farm supply chains.

Opportunities for Livestock Farming

Still, fermentation technology can be an advantage for livestock farming. Animals are more likely to start consuming microbial raw materials before humans will, because precision fermentation produces by-products that also need an outlet. That potential is the subject of a great deal of research.

"This could be interesting both nutritionally and ecologically," says ILVO researcher Annatachja De Grande. "There's a shortage of synthetic amino acids for pig farming. Feed producers are looking for alternatives to imported soy linked to illegal deforestation in the Amazon, and fish feed producers want to move away from fishmeal because of its impact on the seabed. Microbial alternatives might be part of the solution." De Grande, manager of ILVO's new Feed Pilot, underscores the importance of training feed manufacturers. "Today, they work with uniform bulk raw materials. Shifting to microbial by-products will mean using smaller volumes with more variation. That involves more operational complexity. It will be a challenge to quickly assess the feed value of each batch. We aim to build knowledge, run experiments, and guide pioneering companies through the transition."

Of course, much research is still needed, and not just in the lab, but also on pilot and farm scale. Key questions remain: How digestible are these proteins? How do they affect meat or milk yields? Are they safe for animals—and for the humans who consume their products? What impact will they have on nitrogen and methane emissions? And as with all innovation, the aim is to ensure solutions don't simply shift problems elsewhere.

Will Consumers Like It?

The most unpredictable factor in this entire story is the consumer—each of us—and whether we will actually like products made through fermentation. There are signs the alternative protein market is growing worldwide, driven by improvements in taste, quality and price in next-generation meat and fish alternatives. And there are indications consumers are open to it. According to the Good Food Institute (GFI), which has tracked the global alternative protein market for a decade, more than half of Belgian consumers say they would be willing to try cultured meat at least once.

At the same time, a counter-movement is gathering momentum. Italy became the first European country to ban the production and sale of lab-grown meat. Hungary is preparing similar legislation, and the U.S. states of Florida and Alabama have already enacted bans.

Are Europeans Willing to Try Cultured Meat (in Percent)?

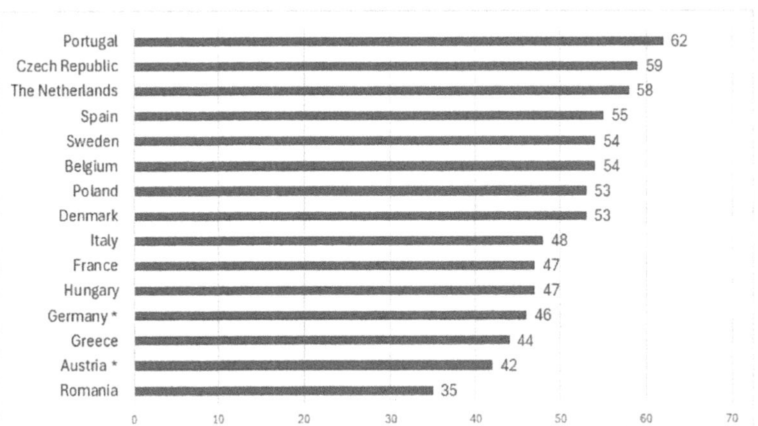

Figure 2: Result of an online survey from YouGov of a representative group of more than 1,000 adults per country, taken in February* or April 2024. Source: *State of the industry: Cultivated meat, seafood and ingredients*, p. 21, GFI, 2024.

The Italian government under Giorgia Meloni justifies the ban as a way to protect Italy's food sovereignty, culture and tradition—a move that may mark the birth of the concept of "cultural sustainability." But the real question is whether consumers will care whether the milk protein in their ready-made lasagna comes from a cow or a microbe—especially if it tastes just as good, costs the same and is just as easy to prepare as any other lasagna on the shelf.

Today, around 70% of supermarket products are already highly processed, and microbial ingredients will belong to that category. This is designer food made from ingredients invisible to the naked eye. One consequence is that we will become even more estranged from our food. When we consume conventional processed foods, we often cannot see what we are eating; but with microbially produced ingredients, even the primary production will be hidden in steel vats in industrial facilities that consumers are unlikely to visit. What's more, the quality, taste, color and shape are always identical. How might that uniformity influence our relationship with food or our attitudes toward food waste? What will it mean for food literacy? Cooking is a skill that is steadily disappearing, which contributes indirectly to the global diabetes epidemic. Microbial "designer food" could help by creating healthy meals tailored to our exact nutritional needs. But it could also make matters worse. How can consumers prepare a balanced, nutritious meal from unfamiliar products? Which way the pendulum swings will depend entirely on how companies use the technology and what kinds of products they choose to develop. And that leads to perhaps the most urgent question about fermentation technology: Who are the producers?

Market Concentration

The agrifood sector differs from most other industries in many ways, but one feature stands out. For decades, buyers and suppliers have consolidated through mergers, while farmers have remained largely independent.

When asked why food companies don't simply buy struggling farms and grow their own raw materials, the CEO of a major frozen-food company answered bluntly, "The risks are too great." Farmers invest heavily in machinery, land and animal housing while their income is directly dependent on the weather and other unpredictable factors. And above all, as we just noted, agricultural products are notoriously volatile—small shortages can send prices soaring, while surpluses can send them plunging.

These risks largely disappear if proteins and fats can be produced by microbes in a controlled industrial environment.

It's no surprise, then, that large food companies such as Cargill are investing in the technology, especially for products such as chocolate, coffee and palm oil, which are becoming scarce and expensive due to climate change or other pressures. The appeal is even greater if production can use the company's own by-products as feedstock or employ already existing brewing kettles. AB InBev hopes to do just that.

Involvement of Companies in Cultured Meat and Dairy Alternatives

	Investment	Acquisition	Partnership	R&D and Production
Cargill	✓		✓	
Coca-Cola	✓			
Danone	✓			
JBS		✓	✓	✓
Maple Leaf	✓			
Nestlé	✓		✓	✓
Nissin Foods Holdings				✓
Tyson	✓			

Figure 3: Analysis by the Good Food Institute (GFI) based on publicly reported events and news in the sector. From *State of the industry: Cultivated meat, seafood and ingredients*, p. 11, GFI, 2024.

The rise of cultured meat and dairy alternatives could further concentrate power in the global food chain in the hands of a few large agrifood players. The so-called ABCD companies—Archer-Daniels Midland, Bunge, Cargill and Dreyfus—already dominate the grain and oilseed markets. Yet that same market power could also drive prices down dramatically. History shows how quickly the cost of new technology can plummet. Take DNA sequencing: in 2000, decoding a human genome cost $1 billion and took 13 years. Now it only takes $1,000 and a few days.

CHAPTER 2

Restore the Soil, Save the World

Just a few kilometers away from where the lab-made cheeses of Those Vegan Cowboys are maturing lies the idyllic village of Hansbeke, a jewel in Flanders' Meetjesland region. A historic castle estate reposes in a lush green landscape. A patchwork of fields and meadows surrounds the stately castle and frames the villagers' picturesque white houses with blue shutters. This environment could hardly be more different from the industrial settings where fermentation technology is taking shape. Yet here too, in this rural idyll, another transformation is underway.

Grain fields are planted alongside faba beans, while grasses are interspersed with herbs and clover. Blooming verges and wooded edges frame the landscape. The fields are plowed less and composted more. And every summer, several hundred farmers, policymakers and researchers gather here to learn about this remarkable phenomenon: regenerative agriculture.

The forces behind this shift are surprisingly similar to those driving fermentation technology, and in a sense, the key actors are the same. Unlike in the 1960s, when organic farming emerged in Europe largely through grassroots ecological pioneers, today's regenerative agriculture movement is advancing from both the bottom up and the top down. Farmers are taking initiative to experiment with soil-restoring practices on their own. Policymakers are embracing the term. Food companies, from multinationals to niche brands, are building it into their sustainability plans. The list of corporate adopters reads like a cross-section of the global food industry. The Swiss company Nestlé, known for snacks and baby food. The American General Mills and Kellogg's, both known for their breakfast cereals. Mondelēz, owner of the popular LU cookies and Côte d'Or chocolate. Dairy firm Danone and its plant-based subsidiary Alpro, canned-vegetable producer Bonduelle, bakery specialist La Lorraine and even IKEA with its famous Swedish meatballs. They all say they have ambitious regenerative agriculture goals and are increasingly requiring their supplier farmers to adopt these practices. They see it as a way to reduce their impact on climate and biodiversity, burnish their public image and comply with Europe's mandatory sustainability reporting (CSRD).

At first glance, this might seem out of step with broader political trends. In the same year as U.S. president Trump announced a withdrawal from the Paris Climate Agreement, the European Commission tweaked the Green Deal and shifted its focus to the Clean Industrial Deal. That says something about current public opinion. Yet outside the political arena, many companies have already spent decades moving along a path toward sustainability, whether through regulation, shareholder pressure or genuine conviction. That momentum is likely to continue.

What Is Regenerative Agriculture?

Unlike organic farming, regenerative agriculture has no formal rulebook or regulatory framework. As a result, not everyone defines it in the same way. In his book *Restoration Agriculture* (2014), which brought the concept to a wide audience, American pioneer Mark Shepard describes a farm that produces food while simultaneously restoring the landscape, soil and biodiversity. As early as 1980, the American Rodale Institute offered a definition that explicitly referred to high yields and good farmer incomes, achieved with minimal environmental impact and minimal use of non-renewable resources. More recent scientific literature defines regenerative agriculture as farming that creates prosperity while restoring soil health, soil biodiversity and the natural cycles of carbon and water. Agroecology is sometimes used as a synonym, but it places stronger emphasis on fairness: fair relationships, fair prices and collaboration throughout the food chain.

Changing Priorities

To understand where this trend comes from, we must look back in time. By the mid-19th century, Belgium was already a major consumer of fertilizers. These ranged from farmyard manure and compost to imported guano—dried seabird and bat droppings from South America—and the first mineral fertilizers, such as superphosphates and metal slag. Government and agricultural researchers of the time embraced the classical fertilization theory (1840) of German chemist Justus von Liebig. His theory did not start from what the soil needed to perform optimally, but what the plant needed to grow: nitrogen (N), phosphorus (P) and potassium (K)—three essential plant nutrients. According to von Liebig, farmers needed to add potassium salts (K) and phosphates (P) to their fields to feed Europe's rapidly growing population. This was no small challenge, given the hunger that followed the failed potato harvests of 1845 and the partial failure of grain crops in 1846.

Evolution of Mineral Fertilizer Use and Agricultural Land Area in Belgium (1880 = 100), 1880-1913

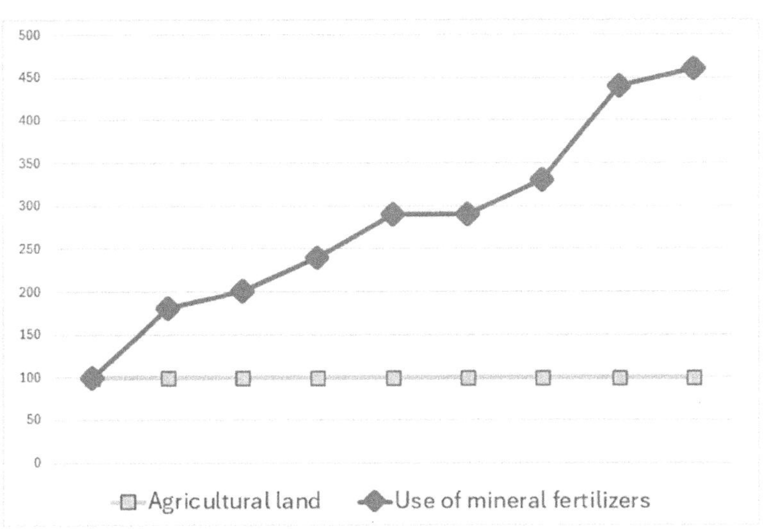

Figure 4: Blomme, Jan, "The economic development of Belgian agriculture: 1880–1980: a quantitative and qualitative analysis," in *Studies in Belgian economic history*, 3, Royal Academy of Sciences, Letters and Fine Arts of Belgium. Economic History Committee, Brussels, 1992.

The mineral salts delivered on their promise. By the early 20th century, Belgian fields were producing record harvests of grain, potatoes and sugar beets.

For decades after World War II, the government and agricultural research once again focused squarely on maximizing yields. The Haber-Bosch process for producing synthetic nitrogen fertilizer, developed in 1910, together with the so-called Green Revolution of the 1960s, gave us the tools we needed. Farmers could access more productive plants and animals, fast-acting fertilizers, chemical crop protection products, machinery and tractors. Flemish agriculture began to specialize in livestock farming, with fewer farms and more animals per barn. Maize (corn) became the dominant field crop. “No more hunger” was a political priority. Regardless of the ensuing environmental and climate side effects, that policy has succeeded. Today, Europe ranks among the world’s largest food producers.

Soil Depletion

The only trouble is that decades of intensive livestock farming driven by fast-acting fertilizers and ever-heavier machinery have taken a toll on the soils. Today, more than 60% of European soils show signs of degradation. Soils in poor condition lose their fertile top layer faster than it can be rebuilt, with compacted layers or surface crusts that block the movement of water, plant roots and vital soil life such as earthworms. Flemish soils are depleted as well. Half of the fields and meadows contain too little “soil organic matter”—referred to throughout this book simply as carbon, its main component. Carbon is a reliable indicator of soil health: it is the basis of fertility and resilience, helps soil resist erosion and boosts the soil’s water-buffering capacity. Soil organic carbon is essential to natural carbon sequestration. When more carbon is stored in soils through photosynthesis, less CO_2 remains in the atmosphere. Unfortunately, in recent decades, carbon stocks in Flemish agricultural soils have fallen, not risen. The reasons are varied: deeper plowing, the conversion of so-called permanent grasslands into arable fields, reliance on synthetic fertilizers and an ever-smaller number of crops in rotation.

Restoring the natural carbon cycle to keep more carbon in the soil and less CO_2 in the air is a core goal of regenerative agriculture.

Carbon Cycle?

Second only to the oceans, soils are the planet's largest carbon storehouse. Fields, permanent grasslands, forests, nature reserves and gardens can capture atmospheric CO_2 through the combined work of plants and soil life. Photosynthesis is key, and soil life also plays a role. Plants draw CO_2 from the air and turn it into sugars to help them grow, releasing oxygen in the process. Some of the carbon they capture ends up in the soil via roots, compounds shed by the roots (exudates), crop residues, compost or the manure of animals that have grazed or fed on the plants. Collectively, this is called organic matter. Soil organisms transform part of this material into nutrients for plants, and part into stable carbon or soil organic matter. As a rule of thumb, every extra ton of stable carbon stored in the soil removes 3.7 tons of CO_2 from the air. But carbon stored in the soil doesn't stay there forever. In temperate climates, roughly 2% of stable carbon is emitted (oxidized) each year by soil organisms. In that process, they release minerals that feed plants (mineralization). And so the cycle continues.

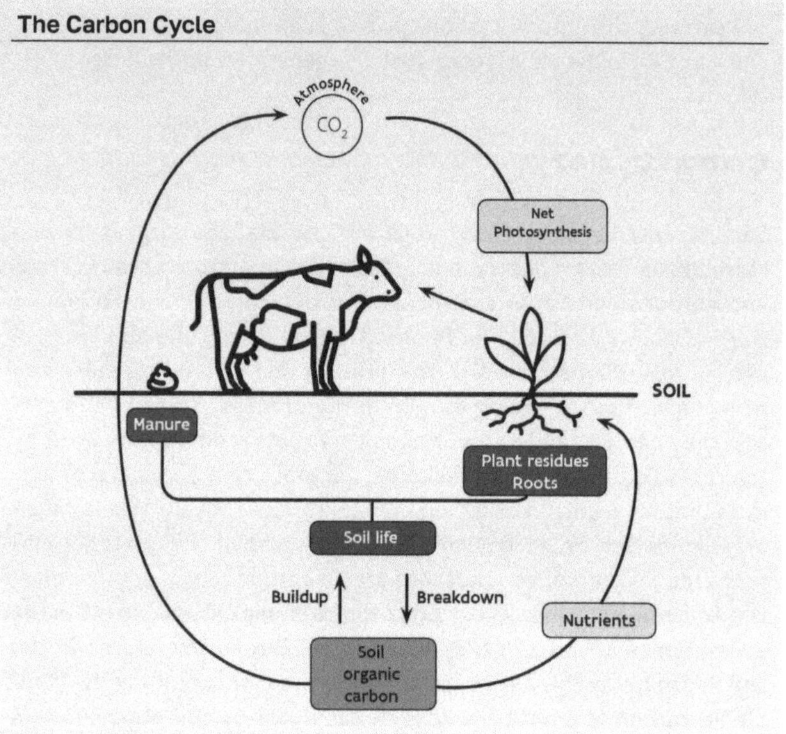

Figure 5: ©ILVO

Evolving Insights

Belgium's grain, potato and sugar beet harvests may still be strong, but uncertainty is increasing—for farmers and their customers alike. Climate change is making the temperate maritime climate of Belgium more erratic, with longer droughts, more frequent heat waves and heavier downpours. Add to that the ongoing decline in soil health, and farmers are noticing that their yields—and their incomes—are shrinking. At the same time, they are facing stricter fertilization limits and an ever-smaller range of available crop protection products. Between 2023 and 2024 alone, the EU banned 14 active substances in plant protection products as part of its plan to cut chemical pesticide use in half by 2030. The post-war formula for agricultural success is under strain. "What will take its place?" asked Flemish professor emeritus Geert

Haesaert at a packed agricultural meeting in Oudenaarde, Belgium in early 2025. His answer was clear: "Technology and innovation can help, but only if our soils are healthy."

Decades of effort are starting to pay off. Green cover crops that protect the soil, capture residual nutrients and suppress pests are now widely adopted. Integrated pest management, which focuses on preventing diseases and pests, is gaining in popularity. And thanks to Flanders' manure policy, the severe over-fertilization of the 1980s and 1990s is a thing of the past.

Is the transition happening fast enough? Reversing the decline in soil carbon is no easy task. At ILVO, there's a saying: "Carbon comes on foot, but leaves on horseback." ILVO soil expert Koen Willekens notes, "Farmers learned a great deal from pioneering organic growers in the early 2000s. Now we're at another turning point—one where farmers are looking outside their bubble. More and more are experimenting with reduced tillage or are exploring on-farm composting."

How Is This Different Than Organic?

Organic farming is a rigorous, legally defined form of restorative agriculture. To market products as organic, farmers must meet strict standards, undergo inspections and hold official certification. Converting to organic isn't easy: one of many hurdles is the required conversion process, which takes at least two to three years. The strict procedure makes it difficult for farmers to get started, but the organic label does offer a quality guarantee for both the farmer and the consumer. In contrast, restorative agriculture or agroecology has no fixed standards, certification or mandatory transition rules. This makes it more accessible for farmers, but also more vague and less transparent. In practice, farms exist along a continuum of sustainability, from simply replacing grass with grass-clover mixtures to sowing clover or a multi-species cover crop in spring while the grain crop is still standing. One key difference: organic farmers are not allowed to use synthetic nitrogen fertilizers and chemical crop protection products, although they may use certain biological or naturally derived alternatives.

The Ten Soil Commandments

Back in 2015, ILVO launched a humorous campaign built around "Ten Soil Commandments." These 10 simple principles for sound soil management were hardly revolutionary, but the campaign showed them in a fresh and surprising light:

1. Thou shalt cherish thy soil above all else
2. Thou shalt descend from your tractor in due season
3. Thou shalt honor the law of eating and being eaten
4. Remember organic matter, to keep it holy and to honor it
5. Thou shalt not drive on wet fields
6. Thou shalt not lay excessive pressure upon the soil
7. Honor the keeping of regular soil analysis
8. Thou shalt not cast fertilizer beyond measure
9. Thou shalt clothe thy soil with green in all seasons
10. Thou shalt rotate thy crops, that thy harvest may be abundant

Long Live Soil Life

Managing the soil as your greatest treasure or capital—that is the essence of regenerative agriculture. Today's scientific movement around regenerative agriculture adds an additional insight to von Liebig's mineral theory, namely that the soil life can also deliver nutrients to plants. You can view soil as von Liebig did: a substrate in which anything can grow, provided you add enough fertilizer and crop protection products. But it is more useful—and more sustainable—to see soil as an ecosystem in which complex processes of eating and being eaten maintain a delicate balance. Here again, microbes and other organisms play the starring role. A single cubic meter of soil can harbor millions of bacteria; thousands of protozoa (single-celled organisms); kilometers of fungal strands called hyphae; hundreds of tiny worms called nematodes; just as many mites, beetles and woodlice; and large numbers of earthworms and centipedes. The richer this soil food web, the greater the chance of getting fertile soil, a good soil structure and resilient crops.

The Soil Food Web

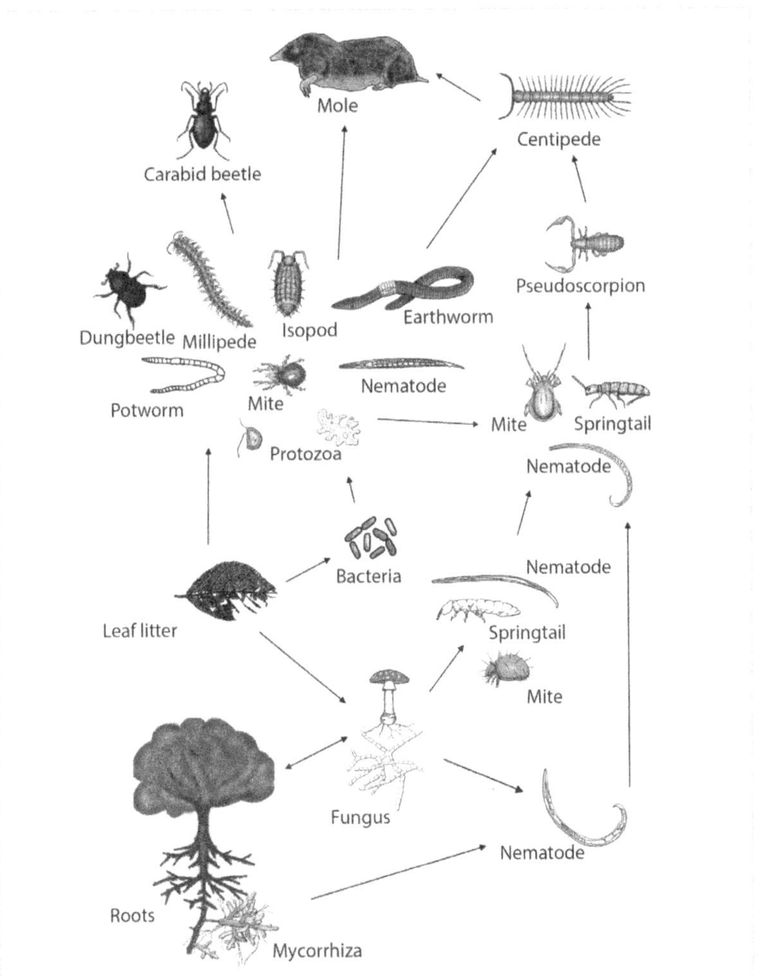

Figure 6: The soil food web, an intricate network of relationships and energy flows among the many organisms that live underground. Original drawing by Ron de Goede, adapted for this book by ILVO.

Soil Structure

In Belgium, the year 2024 was exceptionally wet. Well into spring, large pools of water remained visible on fields and meadows. In a year with normal rainfall, such standing water would be a sign of poor soil structure. Soil consists of solid particles—organic matter and minerals—and the pores between them, which hold either water or air. The way the particles cluster (aggregate), along with the size and volume of pores in different layers, determines the soil's sponge-like ability to store water. The construction and maintenance of soil structure is the work of plant roots and of the wondrous but mostly hidden soil life. Fungi, bacteria and earthworms bind particles into stable clumps with their slime trails, droppings, fungal threads, burrowing and mixing. A healthy soil is marked by stable, crumbly aggregates in the top layer, many pores deeper down and no dense, obstructive layers. A soil with few stable structures, or with compacted layers, plugs up quickly when it rains. This blocks infiltration, causing water to pool on the surface.

Soil Fertility

A healthy soil not only stores nutrients but makes them available to plants. That too is the work of soil organisms that engage in processes of decomposition, symbiosis and weathering at various depths. At the very top, in the upper 10 centimeters, manure and crop residues rich in carbon and nitrogen are broken down by microbes, fungi and bacteria. This creates stable carbon. Gradually, nutrients such as nitrogen and other elements become available to the plants. How well this "decomposition biology" functions depends on soil temperature, moisture and oxygen levels. In cold or waterlogged soils, as in 2024, other, less desirable processes take over.

Deeper down in the root zone, a more complex interplay unfolds. Fungi and bacteria form symbiotic relationships with plants and their roots. They increase the root surface so that plants can absorb less-accessible nutrients. In return for plant-produced sugars (exudates), they even produce defensive compounds against pests and diseases. For this symbiotic process to work, the soil food web needs living plants and requires sunlight to drive photosynthesis.

Then there is weathering—the hardest process to track or measure. Once again, microorganisms are doing the work, extracting nutrients from underground rocks and transporting them through their filaments to plant roots. Humanity owes much to this process: millions of years ago, it enabled plants to leave the sea and colonize land.

Harmful Soil Life

Not all soil organisms are beneficial. Some bacteria, fungi, nematodes and insects cause disease or damage plants. A few hundred fungal species are known to devastate crops when they are present in large numbers. Careful crop rotation usually keeps them in check. But some pathogens are aggressive and sophisticated. One of the most destructive plant bacteria in the world, from the *Ralstonia solanacearum* species complex, can bypass the defenses of hundreds of plant species. It enters soil via contaminated irrigation water and injects proteins directly into plant roots. Eventually, it blocks water transport in the roots, causing plants to wilt, a disease commonly known as brown rot. Immediate action is required and even legally mandated: the brown rot bacterium is a quarantine organism that must be reported and eradicated wherever it is found in Europe.

Recipes for Soil Recovery

Where classical fertilization theory begins with crop requirements and meets those with varying portions of mineral or organic fertilizers, regenerative farming techniques focus first on feeding the soil life. That's easier said than done. Capturing the complexity of all soil processes and relationships in simple "recipes" for soil restoration is nearly impossible. But scientists are trying to gain insight into the tremendous complexity present in the soil. The remarkable farm in Hansbeke, Belgium is a key testing ground. In 2020, ILVO joined farmer Felix de Bousies and advisor Alain Peeters to launch an agroecology trial platform in Hansbeke. Fifty hectares of farmland on a historic estate are being deliberately restored and closely monitored. This unique farmer-scientist partnership has accelerated research into soil-restoring practices in Flanders. It has produced useful insights and has shed light on several key ingredients.

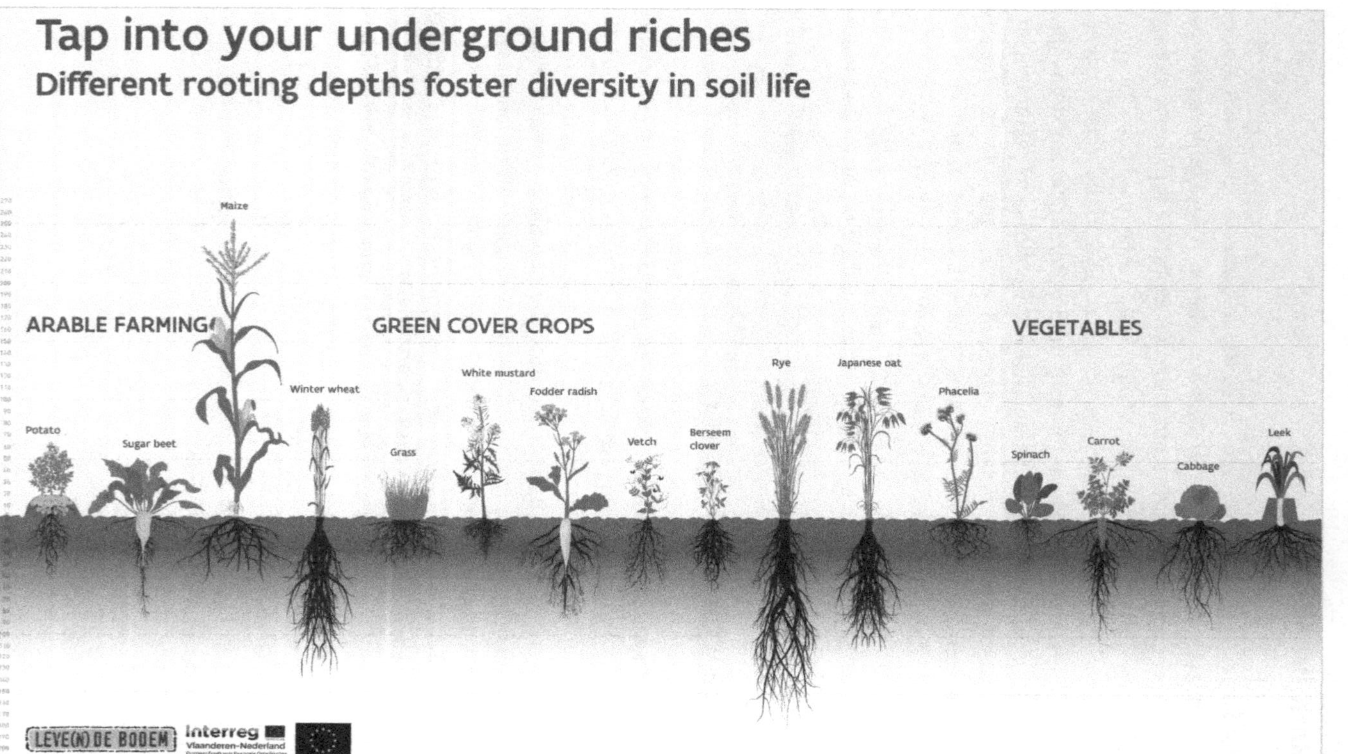

Figure 7: Variation in rooting patterns across a diversity of arable crops, cover crops and vegetables. Source: *Leve(n) de Bodem* (2016–2019). Funded through the Interreg V Flanders–Netherlands program, a cross-border cooperation initiative with financial support from the European Regional Development Fund.

The Value of Plant Diversity

"Diversity is key," says Koen Willekens, coordinator of the Hansbeke platform and a frequent speaker at farmer information sessions in Flanders. "To build a rich soil life, you have to bring as much variety into your crops as possible." This is a tall order, because farmers in Belgium and the Netherlands can't grow crops year-round. The growing season is determined not only by temperature but also by day length. Plants need warmth, moisture and long hours of light to grow. In summer, when light and heat are abundant but water may be scarce, Flemish farmers typically aim to harvest a single main crop—most often grass, cereals, maize or potatoes.

Still, diversifying the crop plan is high on every farmer's agenda. Plants vary in their "thirst" and "hunger" for various elements, they attract different beneficial and harmful organisms above and below ground, and they root in distinct ways. Each crop depletes and enriches the soil in its own way. Some crave soil nitrogen, others less so, and still others even add nitrogen by fixing it from the air. Some crops attract harmful nematodes, while others are natural enemies of the same species. Deep-rooting plants can tap into water reserves during drought, while shallow, wide-rooting plants capture and hold water in a downpour. "For a healthy, fertile soil you need them all," explains ILVO researcher Tommy D'Hose. "Try for as many plant species and families as possible, with the widest possible range of root types. That's what benefits the soil the most."

Mixed Crops

What farmers plant or sow must of course fit into their farm operation. Livestock farmers who produce their own feed—maize, grass and grain—need to harvest enough at the end of the growing season to feed their animals through the winter. They can diversify with mixed crops such as grass with clover or grain with beans. Some pastures are sown with an herb-rich mixture, where grasses and clovers grow alongside species like plantain, yarrow and chicory. This approach doesn't work for potato and vegetable growers, however, as they need to keep their harvest as pure as possible to sell to a processing plant, auction or supermarket. To add variety without mixing crops, farmers can choose the well-established strategy of sowing mixed cover crops.

Cover Crops

Cover crops do exactly what their name suggests: they cover and root into the soil. This is especially important in winter, after harvesting crops like cauliflower or potatoes. Think of the purple fields of phacelia or the bright yellow fields of mustard blooming in autumn. These cover crops are never harvested; instead, they store carbon in the soil. When sown in mixtures, they greatly increase biodiversity both above- and below ground. They improve soil structure and reduce the risk of erosion during wet months. They are also called "green manures" for good reason. "Cover crops play a crucial role in fertilization because they feed the soil life when the main crops have already been harvested. They do this by rooting into the soil and leaving plant residues behind. This is how they keep the engine running for the soil life and help release nutrients from the soil's storage room," explains Koen Willekens.

They can also serve as catch crops, meaning that they "catch" or absorb any leftover nitrogen after maize or potato cultivation, thus preventing it from leaching into waterways as nitrate. They are an indispensable tool for sustainable farming and have been successfully promoted for years. Some farmers are already taking the next step: they sow their main crop directly through the residues of a cover crop, or sow the cover crop through the stubble of the previous harvest. This usually allows earlier sowing, eliminates the need for plowing and limits weed growth.

In Hansbeke, farmer Felix de Bousies is experimenting even further: he sows cover crops even before harvesting the main crop, using a drone to sow the seeds. "It's still very experimental, but the idea is that you can sow even earlier and reduce weed pressure even more. You also use less seed, because the seeds germinate better in the dark, moist environment beneath the crop canopy," says researcher Koen Willekens.

Plants for Carbon Sequestration

Although manicured lawns don't have the best reputation for biodiversity, grass is an exceptionally valuable crop for agricultural soils. Unlike annual crops such as maize, cauliflower, or potatoes, grass stays on the field for long periods, producing large amounts of biomass along the way. This allows it to store substantial carbon in the soil, which is beneficial both for soil life and the climate. Winter grains are also strong carbon-storage crops. They remain in the field for nearly a year.

After harvest, their carbon-rich roots, crop stubble and sometimes their straw all help to replenish the soil's carbon reserves.

Legumes

The agroecological trial platform in Hansbeke sows many cereals, often in combination with a second crop such as faba bean. Not only do the bean and the grain support each other, but faba beans belong to the legume family, a group with special status in regenerative crop plans. Legumes have the unique ability to convert nitrogen from the air into a form they can use. They enjoy a special symbiotic relationship with rhizobium bacteria that live in nodules on their roots. Beans, peas, soy, alfalfa and clover are all leguminous plants. They can be grown as monocultures but also thrive in mixed crops.

Figure 8: Root nodules formed by rhizobium bacteria on a soybean root. Inside these nodules, the bacteria convert nitrogen from the air into a form the soy plant can use. © ILVO

"In mixed cropping with cereals or grasses, legumes can even boost overall yield," explains ILVO expert Mathias Cougnon. "Crops always compete for available soil nutrients. In combinations such as grass with clover or grain with faba beans, the grass or grain wins the race for soil nitrogen. This forces the clover or faba bean to obtain nitrogen from the air by investing in its relationship with root bacteria—a costly process, since the bacteria demand sugars in return. The result? The grass or the grains get their full portion of soil nitrogen, and the leguminous clover or faba bean also perform better than they would in monoculture. Here, one plus one really is more than two."

The Value of Fertilization

It's now well known that nitrogen and phosphorus from manure can leach into waterways in the form of nitrate and phosphate, harming biodiversity. Strict rules govern how much and when farmers may spread manure. Fertilizer recommendations are based on soil analyses, and enforcement is strict. But weather affects nutrient uptake, and soil life can store nutrients temporarily and release them later. Fertilization is therefore a balancing act, and many farmers prefer mineral fertilizers because of the precise dosage. Organic fertilizers such as farmyard manure, slurry and compost are more variable in composition, and their nutrient release is also less predictable. Dosage with organic fertilizers appears to be more difficult.

"Diversity matters here, too," Willekens stresses. "Plants don't only need nitrogen, phosphorus and potassium to grow and remain healthy, but also minerals like calcium, magnesium, iron, manganese and zinc. Organic fertilizers contain all of these." This means that while mineral fertilizers can sometimes be used as a supplement, the main fertilizer should come from high-quality animal manure, farm compost, woody residues from nature management and biomass from green manures, depending on what's available at the time. "Organic fertilizers feed soil life and supply not just nitrogen but also carbon. This reduces the risk of nitrogen leaching into the environment and ensures fertilization does what it should: replenish the nutrients removed with the harvest and keep the soil's carbon pantry well stocked."

For successful natural fertilization, farmers need more knowledge and experience in practice, as well as reliable manure quality and enough biomass for compost. Can we farm without mineral fertilizer? Not without yield loss. Can we farm without animal manure? Not without long-term declines in soil quality and productivity. Manure may have a bad reputation due to past overuse, but nicely matured manure that is rich in carbon and microbial life is invaluable for soil health. Its reputation should be rehabilitated. Some initiatives are already working on that, such as the "Top Manure" contest held in the Netherlands and Belgium. Agricultural advisor Peter Vanhoof celebrates dairy and pig farmers who produce the best quality manure, defined as being low in nitrogen emissions and highly nourishing for soil life. Much attention is given to the feed that the winning farmers give their animals, because this strongly influences manure quality. Participants learn from each other while manure and animal feed both receive the recognition they deserve.

On-Farm Composting

On-farm composting used to be standard practice in agriculture, but it has fallen out of favor with the rise of chemical fertilizers and the overabundance of livestock manure. The compost you make at home is probably made from kitchen and garden waste. By using only leaves, clippings, or pruning waste, you get leaf or green compost. Farmers add livestock manure to the mix. This mix of green and "brown" materials decomposes into a carbon-rich soil conditioner applied to the fields in autumn. To produce high-quality compost, farmers need a large and diverse supply of biomass—something usually not available on a single farm. That scarcity is one of the main obstacles to wider adoption of on-farm composting. In the cross-border Saeftinghe nature park between Belgium and the Netherlands, farmers and landscape managers are now experimenting with collective composting to overcome that hurdle.

Minimal Tillage

The plow has played a key role in the development of agriculture. Especially in regions with a heavy clay soil, like Flanders and the Netherlands, plowing brought fertile harvests and prosperity to the people. Today, however, plowing is under discussion and more and more farmers are experimenting with shallower, less disruptive ways of working their soil.

"A plow breaks up and turns over the top 30 centimeters of soil—the zone where young plants establish," explains Tommy D'Hose. "It buries weed seeds and mixes crop residues as well as fertilizers into the soil layer, distributing nutrients and carbon more evenly."

Those are good arguments for plowing, so why the debate? Plowing inevitably damages soil life and in recent decades farmers have plowed deeper and deeper, with ever-heavier machinery. This has depleted soil carbon stocks and degraded soil structure. Shallow, non-inversion tillage reduces that damage and helps keep more stable carbon in the topsoil, which in turn reduces erosion and waterlogging.

"There are many caveats," D'Hose stresses. "Sometimes plowing is necessary, but sometimes it's better not to. Abandoning plowing altogether is not the holy grail of soil restoration. What matters most is choosing the right timing and frequency."

Technologically, it is entirely possible to work the soil less deeply and less intensively. Farm mechanization has come a long way since the tractor and plow became standard tools in the 1950s. Today, direct seeders can crush a cover crop and open only a narrow slit for sowing seeds. There are plows that invert less than 10 centimeters of soil, cultivators that precisely skim off a sod or cover crop and shallow rotavators that slice off the grass sod without disturbing the soil underneath. These machines exist, but are they widely accessible? "That's the sticking point," says Koen Willekens. "There's a real need for skilled contractors who invest in this machinery and know how to operate it properly. That's why we're seeing local initiatives to create machinery cooperatives, where farmers share not only the equipment but also the services of trained operators."

What About Yields?

It's not true that regenerative farming yields less than conventional farming. In fact, there is no single definition of "conventional" or "regenerative" farming. Both exist on a spectrum, as do the farms.

ILVO's experimental fields have already generated top yields in organic cabbages. And Belgian organic farming pioneer Jan Van Overbeke grows healthy, high-yield crops on his farm in France. "We see a knowledge gap among farmers and researchers alike", says Koen Willekens. "There is a lack of experience and there aren't enough suitable crop varieties, specialized contractors, biomass for on-farm compost and so on. We're still pioneering. The recipe book is still being developed. But step by step, we're catching up." Felix de Bousies, owner of the innovative farm in Hansbeke, adds, "Of course it has to be profitable. I get nice grain yields—maybe not the record yields of winter wheat or barley—but I still earn more. In part because I save on pesticides and fertilizers, but mostly because I can negotiate good prices with the millers who buy my wheat. For me, it's simple math: the operation needs to stay profitable."

An Industrial Lever

There is no exact count of how many farmers are practicing regenerative agriculture. The best we can do is gauge adoption based on its key ingredients. Crop rotation? Increasingly common. Cover crops? Now mainstream. Reduced tillage? Still niche, but growing.

It's too early to speak of a breakthrough, but the seeds are clearly sprouting. In just six years, attendance at the annual agroecology demonstration day in Hansbeke has tripled. And it's not just ecologists or organic growers who show up to see the trial fields. The audience is diverse: farmers, advisors and consultants, all eager to learn. How much did the crops yield? How well does faba bean perform in cereal mixtures? Is there profit in buckwheat? These are entrepreneurs navigating the boundaries set by policy and the free market. They come out of curiosity, but also because policymakers and downstream buyers—retailers, food processors—are beginning to set requirements.

"There is pressure from Europe via the Green Deal and the Common Agricultural Policy, but also through the Corporate Sustainability Reporting Directive (CSRD)," explains ILVO sustainability expert Veerle Van linden. "Large food companies must report on their sustainability efforts, and that extends across the entire value chain—from raw materials to packaging to the by-products that end up in the trash."

According to *Food Navigator* (2025), a European food industry journal, security of supply is another driver. In a warming climate, food companies expect that regenerative systems will offer more stable harvests. They also count on "carbon farming" to offset part of their own CO_2 emissions. Yet since neither regenerative agriculture nor carbon farming has a strict definition, each company creates its own interpretation.

"It's not always clear whether these programs deliver what companies claim," Van linden cautions. "But their influence—and potential leverage on farming—is enormous. We've already seen it in dairy, with requirements around grazing and climate audits. Participation first comes with a bonus, then it becomes the new baseline. Policy can point the way, but in practice, it's the dairies, cookie factories, potato processors and supermarkets that use financial incentives and market power to steer farming in a certain direction."

Mondelēz was one of the first food giants to impose conditions on its suppliers. Since 2008, through its Harmony program, it has worked with wheat farmers near its cookie factories in Europe. In exchange for a better price, farmers adopted biodiversity-friendly practices on their fields. Today, 1,100 farmers across seven European countries (including Belgium) participate, together producing about 400,000 tons of wheat—almost all of the wheat Mondelēz Europe needs to make its cookies. By 2030, the company wants to reach 100%. Competitors Nestlé and Danone joined later but are catching up fast: according to *Food Navigator* (2025), they already source 21% and 39% of their raw materials, respectively, from "regenerative" farming.

The Potential of Carbon Farming

Through the soil-restoring practices described earlier, farmers can indeed raise the carbon content of their soils. The term "carbon farming" has thus entered our lexicon. But whether these measures are enough to offset the greenhouse gas emissions of multinationals like Mondelēz, Nestlé, or Danone is another matter. The fact is that European agricultural soils are still losing carbon today. In absolute terms, that loss amounts to 25.7 million tons of CO_2 per year. Halting that decline is the first priority before we can even start talking about net carbon sequestration. And carbon is not the only greenhouse gas involved. Nitrous oxide, a powerful greenhouse gas, can also escape from soils. To truly understand the climate benefit of carbon-building measures, we must understand and account for both processes.

Soil carbon storage is not unlimited either, and not all soils have the same capacity. Clay-rich or wetter soils can store more. Protecting existing carbon hotspots such as permanent grassland is far more important for the climate than adding compost to cropland or sowing extra cover crops. "Sequestering carbon in the soil is a gradual process," says ILVO soil scientist Greet Ruysschaert. "Various studies show that it can only compensate for a portion of the farm's own greenhouse gas emissions.

Despite the many benefits of carbon farming, we should be careful not to overstate its potential to slow down global warming."

Carbon Credits

Financial compensation is available to farmers if they make an effort to store additional carbon in the soil. Both public and commercial initiatives exist, but not every initiative is equally transparent. That is why Europe is working on standards and a harmonized system. In Flanders, a Flemish Action Platform on Carbon Removals and Carbon Farming was recently launched. Clearly, things are moving. But before carbon farming can become a reliable additional income stream for farmers, more work is needed. ILVO researchers Ennio Facq and Jeroen De Waegemaeker issued a policy brief with recommendations on this topic in 2025.

Agroforestry

Often mentioned in the context of carbon farming is agroforestry or food forests. Here, crop or livestock production is combined with trees that provide timber, fruit or nuts. For carbon storage, the system offers a double benefit: carbon is sequestered not only in soil but also in long-lived trees and shrubs. Belgium currently counts only about 100 agroforestry farmers. In the Netherlands, the practice started later but is expanding more rapidly.

Crop Protection

Carbon storage is not the only area where claims and expectations from regenerative agriculture sometimes run ahead of reality. Nearly mythical claims abound online: fields that fertilize themselves without any external input, crops that thrive without any crop protection at all, soils that single-handedly stop climate change. *Restore the soil, save the world?*

"Regenerative agriculture is not nature—it's still agriculture, where the point is to grow productive crops," stresses ILVO researcher Koen Willekens. "The idea is to make maximum use of natural processes: you feed the soil life that in turn feeds your plants, and you attract a limited number of pests so their natural enemies can also establish and prevent further spread. But there are challenges, and there are still gaps in our knowledge. I'm sure we will come up with solutions, but currently weeds and pests can be difficult to control."

Work is ongoing on a wide range of alternatives. Natural crop protection products are being developed, along with bacteriophages (tiny viruses that attack plant-pathogenic bacteria) and microbial inoculants such as compost teas that strengthen plant resilience. Mechanical solutions are becoming ever more precise: machines are being developed to hoe, pull or burn weeds. Artificial intelligence is being applied to improve monitoring and early-warning systems. Breeding programs are delivering more robust crop varieties. And insights are evolving about the design of resilient food systems. Field trials in Hansbeke that test the planting of rough field margins to reduce insect pests or the use of mulches and crop mixtures to suppress weeds have shown mixed results. On the ILVO experimental fields, researchers use robots, drones and AI-supported tools, but clearly, these technologies are not yet accessible to or affordable for every farmer.

One thing is certain: the chemical crop-protection products that are rapidly disappearing from the market cannot be replaced one-to-one. "What we need is an integrated approach," explains ILVO spray-technology expert David Nuyttens. "Interventions—chemical or otherwise—are not taboo, but they should be the last step. Prevention comes first. And if that is not enough, new innovations can help us apply small, precise doses of selective yet effective products exactly where they are needed."

Entrepreneurship

"How much damage are we willing to accept?" is a philosophical-economic question sometimes raised in the context of crop protection. But it overlooks one key fact: farmers are entrepreneurs. Each one must find their own balance—often in precarious circumstances. Consider other food producing regions in Europe: deserts are advancing north into Spain, Italy and Greece. "In our fertile river delta we are not confronted with desertification. So far, we have been spared the worst effects of climate change. We haven't had any dramatic harvest failures for a long time. This creates the illusion that affordable food will always be available, everywhere, for everyone. But we too face risks that must be managed with the best available techniques. We cannot afford to rule out tools in advance," stresses Isabel Roldán-Ruiz, scientific director of ILVO's Plant Sciences department. She points to the potential of precision farming using AI and robotics, as well as the breeding of more robust crop varieties and the promise of new genetic technologies. We'll explore both topics in later chapters. But first, let's play a short game.

Sit up straight and use your imagination.

Picture a farm practicing regenerative agriculture. Is it large or small? Are there animals? A shop? A terrace? Can anyone walk in, and who would you meet there?

Sorry, dear reader, but we cheated a bit there. Our questions probably nudged you in the direction of imagining a small, nostalgic farm with chickens scratching near the playground, pigs rooting in straw, a small tractor in the barn and a friendly farmer or farmer's wife serving you in the farm shop. You buy a few leeks and a block of cheese. Maybe you join a quick tour of the cow barn with your children or grandchildren, and enjoy a satisfying ice cream (*please, can we have one?*) on the terrace before you leave.

The reality? Fewer and fewer people today have grown up on a farm and know how a farm actually operates. The idyllic picture above still exists, but only on about 2% of the farms in Flanders. They are clearly the exception, because it's not an easy way to make a living. Regenerative agriculture isn't either. The soil restoration practices described in this chapter demand either more labor, new (often costly) technologies or most likely both.

"Environmental requirements, including those we now translate into regenerative agriculture, are making farming more complex, more labor-intensive and more capital-intensive. The farmer must show even greater entrepreneurship, professionalism and resilience," says Fleur Marchand, scientific director of ILVO's Social Sciences department. "The question is which type of farm or farmer can make this a reality." Two theoretical models are often mentioned for farmers making this transition: smaller farms that sell exclusive products to restaurants, hotels or specialty shops at high prices, or larger farms that produce for export at lower prices but can invest in automation and technology due to economies of scale. Both models can apply regenerative practices, regardless of their size. These are only models—in reality every farm will be a hybrid—but these models highlight a key turning point in agriculture. What about the average farm? How do we make space for farming in an urbanized Flanders where housing, industry and nature all compete for land? How do we support farmers who can't make the transition? And how do we build fairer food chains?

"These are difficult questions that cannot be solved by science and technology alone," says Marchand. Or as Flemish organic farmer Bavo Verwimp wrote in a 2022 op-ed in Belgian newspaper *De Standaard*: "If we truly want solutions for food production, we also need sociologists, psychologists and economists."

Fair Chains?

One question for those economists and sociologists: What constitutes a fair income for farmers? The topic rose high on the agenda after the 2024 farmer protests in Flanders and the Netherlands. Interestingly, farm income was not the main driver of discontent, nor is it directly linked to supermarket prices. "Farm income is strongly tied to global markets, where supply and demand play the dominant role. Small shortages cause big price spikes, and vice versa. But these swings are not always reflected in consumer prices. If the wheat harvest fails in Romania, our bread doesn't suddenly become more expensive. The shock is absorbed and diluted along the long food chain," explains ILVO agricultural economist Erwin Wauters.

Calls for "fair prices" or "true prices" demand transparency—not only about who absorbs price shocks, but also about the hidden social and environmental costs throughout the chain. Current prices conceal low wages in the Global South, possibly unsafe working conditions in factories or ports, biodiversity loss from polluted water, greenhouse gas emissions and the health impacts of air pollution. In a recent policy brief, ILVO concluded that current market prices fuel social inequality and reward unsustainable production. "Excluding costs for nature, health and society from the food chain makes it more profitable for companies to invest in unhealthy, unsustainable products. In other words, the companies that excel at externalizing costs attract the most capital. Meanwhile, the most vulnerable communities bear the brunt of the impact," says ILVO researcher Valentina Martinez Ruiz. Despite the importance and the urgency of this societal question, few concrete solutions are being presented. Experiments in supermarkets with variable pricing—displaying both the shelf price and the "true cost"—have shown mixed results.

On a smaller scale, there are hopeful examples of chain partners that set prices together, thus ensuring a fair income for both sides. For example, farmer Felix de Bousies sells organic wheat to millers and bakers at a hectare-based price—essentially a performance-based contract rather than an output-based one. This ensures risk-sharing across the chain. "Sometimes partnerships appear in unexpected places," notes Dylan Feyaerts, coordinator of ILVO's Living Lab for Agroecology and Organic Farming. For nearly a decade, hospital group AZ Zeno in Knokke-Heist has sourced fresh vegetables from the nearby CSA farm Het Polderveld. The guaranteed outlet provides stability to the small farm, while the hospital reduces its catering costs by buying locally grown produce.

"These market-driven initiatives are encouraging," continues Feyaerts. "But for a real transition, policy support is essential—with real instruments and clear goals. There are examples of farmers leasing public land for regenerative practices, contracting with authorities to protect farmland birds or to supply ingredients for affordable school meals. But these are ad hoc cases, not part of a wider vision. Making regenerative farming a success will depend on solving the problem of access to farmland, resolving compost shortages, finding ways to promote peer learning and creating mechanisms to make fair pricing possible."

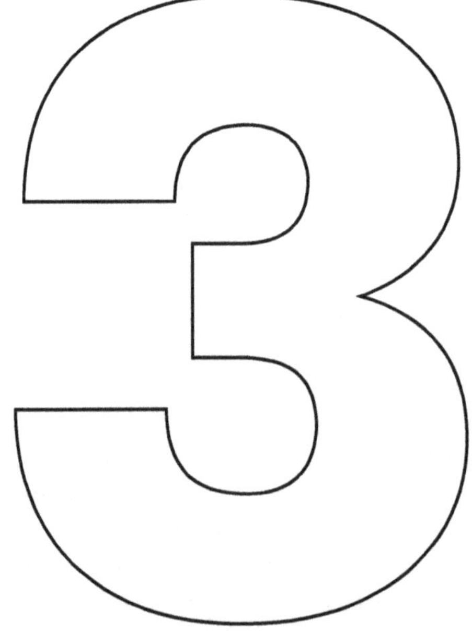

CHAPTER 3
Artificial Intelligence on Land, at Sea and on Our Plate

In the dry dock at Zeebrugge on the Belgian coast, a vessel rests on blocks, exposing its blue-painted hull. The Z26 *Avanti* has gotten a wash and is now undergoing maintenance. On the dock around the massive propeller, a small crowd has gathered: journalists, scientists, politicians, chairs and directors, the shipowner and his family. This afternoon in late spring 2024 is a notable one: this ship is the first in the Belgian fleet to be fitted with new sensors on its hull and propeller. A Doppler speed log and a torque meter will provide unprecedented insight into propulsion and fuel use during navigation and fishing. "Another step toward more climate- and environmentally conscious fishing for Belgium, and a European first," says Hans Polet, scientific director of ILVO's Marine department.

A year earlier, on an ILVO field in Merelbeke-Melle, Belgium, a different first was on display. An electrically powered, fully automatic tractor was preparing a seedbed for savoy cabbages without a driver in sight. This "Djust-e" tractor and the harrow it pulled were operating completely autonomously. The relief on the engineers' faces was unmistakable. "This is a turning point. The impact on farming as we know it cannot be overstated," beamed Jürgen Vangeyte, scientific director at ILVO's Technology and Food Science department.

The institute has decades of experience in researching precision techniques for agriculture, fisheries and food. For a long time, however, the cost-benefit balance was uncertain: systems were complex, often not aligned enough with the real needs of farmers and difficult to integrate into daily practice. "With the rise of artificial intelligence, improved data structures and sensor technology, we can finally master that complexity, and movement is happening in the field," says Vangeyte. We are only at the start of what may become a true data revolution, with far-reaching implications for how we fish, farm, process food and safeguard food safety.

Data Farming in Space

The digital world may feel far away when you're on vacation, buying fresh fish right off a boat, or when you're treating yourself to a box of strawberries from the market on Saturday. And yet farmers and fishermen generate vast amounts of data. Milking robots automatically record how much milk each cow produces, potato harvesters log yields per field and shipowners on shore know in real time how much fuel their vessels are consuming out at sea. "Data is everywhere. But we have to harvest and share that data fairly if we are all to reap the benefits," notes Vangeyte. ILVO's Technology and Food Science department is a European leader in this area. In 2019, ILVO researchers launched the Flemish agridata space called "DjustConnect" which now links one in five Flemish farmers with various public authorities and numerous companies in the fast-growing digital ecosystem. Legally, ethically, technologically and economically, DjustConnect is seen as a model for other EU member states. "The farmers remain in control, so the added value of the data generated on their farm flows back to them. There are clear guarantees that everyone will abide by the rules. And because the network is local, all apps and tools developed with farm data are genuinely relevant and useful for them," explains coordinator Stephanie Van Weyenberg (ILVO).

In 2023, ILVO and its European partners also founded the AgrifoodTEF: a Test and Experiment Facility (TEF) for testing and validating innovative data, AI and robotics solutions for the agrifood sector. The testing environment is safe and controlled yet realistic. "Start-ups and companies from across Europe can give their robots, drones, data applications or AI models a test run here in practice. That accelerates and streamlines their route to the market," says TEF manager Marijke Hunninck.

Since 2025, ILVO has also taken the lead in building the Common European Agricultural Data Space. "This data space is the final piece: connecting existing and new European initiatives into one 'federated' system. This allows all European actors—farmers, businesses, governments, researchers and even consumers—to share data securely, transparently and with ease," says Vangeyte. "Based on that data, true added value can be created, not least by simplifying administrative processes, but especially by developing a powerful new generation of AI-driven applications."

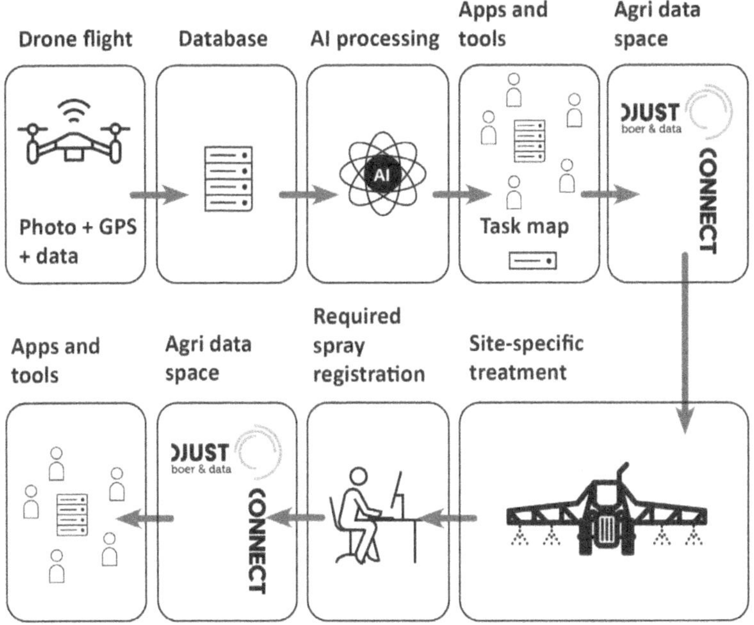

Figure 9: Example of an existing data flow. A drone captures images of weed pressure on a field. Using artificial intelligence, this drone data is processed into a site-specific task map for a sprayer, which is uploaded into an agridata space such as DjustConnect. The sprayer reads the task map and applies treatment only where needed. Data from the sprayer can be uploaded into the digital spray register. Via the data space, it then becomes available for apps and tools used by the farmer. © ILVO

The 14 interconnected data spaces should help Europe catch up in the field of artificial intelligence. Today, too much data remains fragmented, unused or beyond European control. "As long as the costs stay high, the systems remain complex and the connections are missing, we are failing to unlock the potential AI has to offer. It's high time that we took control," Vangeyte argues.

So what is AI's great promise for agriculture and food? "Working with living systems like animals, plants and microorganisms is extremely complex: they grow, they respond to their environment, and no individual is exactly the same. Classical predictive models quickly reach their limits. Artificial intelligence fed with a diversity of current and practice-relevant data—precisely what the European agridata space enables—is far better at handling variability and uncertainty, and learns much faster. This paves the way for robust, practical models for applications in precision farming and precision food production."

At the same time, AI is making precision technologies more accessible for farmers. "Today, precision farming is often too complex: there are too many separate apps and tools. With AI, you'll simply be able to speak to your tractor and ask it to fertilize your field based on last year's yield data and soil records." The biggest leap forward, according to Jürgen Vangeyte, will come with the arrival of multimodal AI models. These systems can process multiple types of data at once: satellite and drone imagery, soil sensors, weather data, text-based advice, video tutorials, legislation and even real-time market information. "They'll be able to interpret complex farming situations in their local context and provide farmers with useful, precise suggestions. This means quicker and more efficient decisions, without being overwhelmed by the flood of data 'under the hood.'"

For global food production, this heralds the prospect of higher yields with a smaller environmental footprint. More food with less fertilizer, smaller amounts of crop protection products, less water, fewer antibiotics, less land and less time.

Fishing for Data

At sea, developments are moving just as quickly. Belgium has a special place in this story. The unique, robust data network that covers nearly the entire Belgian fleet–VISTools–was the brainchild of a Flemish skipper. For years, Pedro Rappé watched with frustration as a black box in his wheelhouse collected information for European monitoring, while he himself had no access to the data. He built a rudimentary computer to tap into the dozens of sensors onboard and eventually turned up at ILVO's marine research department with a box full of cables. The rest is history. Today, VISTools shares data not only with the skipper and crew at sea, but also with shipowners and scientists on shore.

The Belgian Fishery

With just 64 commercial vessels, Belgium has a relatively small fleet. Most use beam trawls and other towed gear to catch flatfish such as plaice and sole. In recent years, the sector has replaced many of its older vessels with newer ones that are far more up-to-date and energy-efficient. Belgian fishermen operate not only along the Belgian coast, but also across the North Sea and the Eastern English Channel, reaching as far as the Irish Sea and the Bay of Biscay. ILVO's marine researchers in Ostend study sustainable fishing techniques, monitor the ecological impact of bottom trawling and gather the data needed for scientific stock assessments.

What Rappé understood early on was the enormous potential of data-sharing, both for individual fishermen and for the long-term profitability of the entire sector. Fisheries are a strategic sector for Europe's food autonomy and economic security. But fishermen need to catch new fish every year. After filling their nets, they need to leave enough mature fish in the sea to renew and sustain the population. Since the early 20th century, ICES[1], an international network of scientists, has been collecting data from different countries to monitor fish stocks in the North Atlantic. That process has remained unchanged for more than 100 years: people catch fish using nets and then count, weigh and measure each individual in the nets. After analysis and modeling, they produce estimates of stock sizes and the associated risk of overfishing. Governments use those estimates to set annual catch limits and quotas per species. The limits are then applied collectively.

---

[1] ICES is the International Council for the Exploration of the Sea, a scientific council that provides independent advice to national and international governments. Approximately 1,000 marine scientists from research institutes and universities are affiliated with this council. ICES operates in the North Atlantic Ocean, including the Baltic Sea and the North Sea.

Data Collection On Board a Fishing Vessel

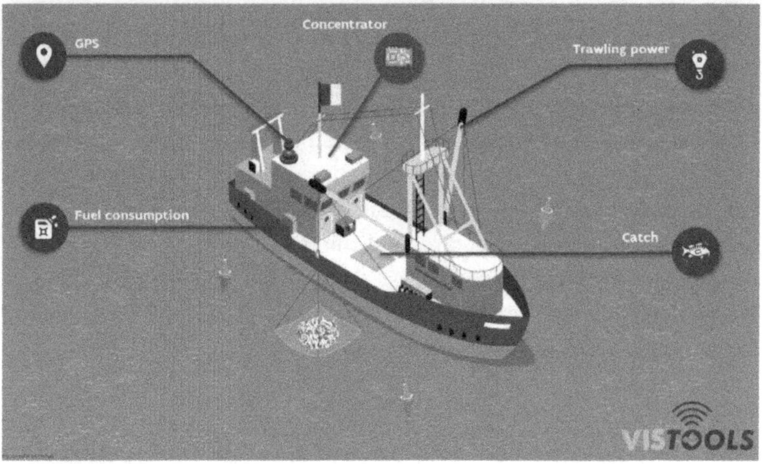

Figure 10: The VISTools data network collects information on catches, fuel use, towing power and location from Belgian vessels. That data is gathered aboard the vessel and shared with the shipowner and researchers on shore.

“The human effort behind these quotas is important, and it will always be needed. But in reality, our current methods only allow us to sample 2% of all catches. It’s an expensive, time-consuming process, with risks of incomplete or inaccurate data. VISTools, artificial intelligence and new DNA techniques now allow us to collect much more data from an unlimited number of vessels, fishing grounds and even species that aren’t currently monitored. We can analyze them faster, too, enabling both fishermen and authorities to respond more quickly,” says Hans Polet. “Concretely, this means quotas would no longer be based on observations from two years ago, as is the case today, but adjusted continuously on the basis of real-time data. It’s clearly a win-win.”

How Are Fishing Quotas Set?

Like agriculture, fisheries are a core European policy area. Each year, the EU Council of Fisheries Ministers determines the maximum quantity of each species that may be caught in the following year. They based this decision on scientific advice from ICES, which draws on systematic data collection and analysis by institutes such as ILVO. Each member state then receives a national quota, allocated according to a historically agreed distribution key. National or regional authorities are responsible for ensuring that their fleets respect these quotas. The goal of the EU fisheries policy is the sustainable exploitation of fish stocks. For stocks beyond European jurisdiction, international agreements apply.

eDNA from Seawater

Imagine not needing to catch a single fish to know what's swimming in the sea. You just take a water sample while sailing, and a DNA device tells you which species are present beneath the waves—and in what numbers.

Now imagine if every vessel at sea did the same, or if fixed sampling points were placed on offshore structures like wind turbines.

The ocean could transform overnight from one of the most data-poor environments on Earth into one of the richest. "We're not there yet," cautions Hans Polet. "The technology already works at lab scale. Now we're testing whether it also works at sea." Together with a Portuguese partner, ILVO has built a prototype device that can be installed on deck to automatically collect, label and preserve water samples—with no human intervention required during the voyage. What it cannot do (yet) is analyze those samples onboard. "The samples still need to be sent to a laboratory," explains ILVO researcher Sofie Derycke, a specialist in environmental DNA, or eDNA for short. "In the lab, we screen the DNA found in each sample and compare it with scientifically validated databases. If there's a match, we know with certainty that this fish, lobster, worm, snail or squid lives within a five-kilometer radius of the sampling site." *Lives*—in the present tense—because the animals do not need to be caught.

This technology provides a non-invasive method to monitor marine species and estimate the biomass present. Does it find one stray individual at a time, or an entire population? "This information complements traditional catch monitoring, and it reveals even the tiniest organisms that live in the water but never show up in fishing nets," says Derycke.

Cameras that Count Fish

Onboard catch monitoring is also becoming more accurate thanks to cameras that automatically count fish as they pass by on sorting belts. This is already happening at pilot scale. Remarkably, the models analyzing the images are trained not only on real photos, but also on synthetic data: lifelike, AI-generated images. “Our programmers can now distinguish even very similar fish species,” says Jochen Van Lysebettens, spokesperson for the West Flanders AI company Vintecc. On their computer screens, thousands of AI-generated images of different fish species have flashed by, in every size, from every angle, with and without tails, with or without skin lesions, stacked neatly or buried under other fish. “That’s the challenge on board a boat: fish never pose politely for the camera, and they never look like their ‘passport picture’. With AI, we can build an extensive portfolio of how each species might appear on a conveyor belt.”

Today, the model used at ILVO can already identify 19 species correctly. “We don’t intend to replace people counting fish on board,” stresses ILVO scientist Sander Delacauw. “Human validation will always be needed. Just like a scale, the models must be recalibrated regularly to prevent bias creeping in. Our aim is to use the cameras to monitor far more than the current 2% of catches. Currently the cameras only go on board with our observers, and we see that they deliver much richer, more detailed data for each voyage.”

Reading Otoliths

At the InnovOcean Campus in Ostend, home to ILVO’s Marine department, a technician deftly removes tiny bones from behind a fish’s gills. Put these “otoliths”—ear stones—under a microscope and you will see growth rings, much like the rings of a tree. You can use them to determine a fish’s age, but only if you have the eyes of an expert. “Otolith reading” is a highly specialized branch of marine biology that takes years to master. Fortunately, ILVO began building a database of otolith images and coupled them to their peer-reviewed ages many years ago. Today, this database is being used and expanded by research partners around the globe.

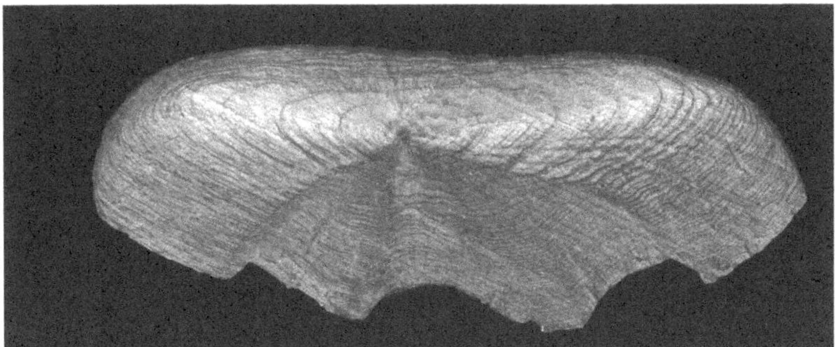

Figure 11: Cross-section of a sole otolith. © ILVO

"This is a gold mine of high-quality data that we now use to train an AI model. That model can either support human otolith readers in the lab or provide a second opinion in case of doubt," explains AI developer Sam Vanhoorne (ILVO). Why does this matter? "At sea, our observers count and measure fish, but knowing their exact age is important to assess the health of fish populations. If too few young fish are caught, that's a warning sign that the stock is in trouble. By giving otolith readers an AI assistant, their work becomes faster and more accurate. That leads to better stock assessments, more precise catch quotas and ultimately a healthier marine ecosystem."

Vessel of the Future

Back to the Z26 *Avanti* suspended in the Zeebrugge dry dock in May 2024. Shipowner Steve Depaepe demonstrates on his smartphone how he can track his vessel's fuel use in real time with the VISTools app. His *Avanti* is a relatively new ship, already far more fuel-efficient than her predecessor, but the app helps his crew save even more. "Fuel accounts for 20% to 30% of a skipper's costs. Even small differences add up," says Sander Meyns, shipowner representative. The speed and torque sensor installed on the *Avanti* and several other vessels in 2024 is helping to further refine the research into fuel savings. "It gives us insight into the factors that drive energy use. We can see where the gains are, both big and small, and shipowners can benchmark their consumption against similar vessels. This way, Belgian fisheries can cut energy use already in the short term," Meyns explains.

Looking further ahead, these insights also open the door to the development of vessels that are even more energy-efficient and rely minimally or not at all on fossil fuels. "The data from VISTools, combined with predictive models, enables us to build digital twins," says ILVO researcher Lancelot Blondeel. A digital twin is a virtual, interactive replica of reality—in this case, a fishing vessel and its operations. It allows researchers not only to study vessels in detail but also to test various design or policy scenarios: "What happens to shape, weight, energy use and autonomy if you replace a diesel engine with a larger hydrogen engine? How much space remains in the hold for fish? The digital twin immediately provides feedback when you adjust the parameters, turning a complex theoretical exercise into a set of rich, detailed and useful images."

Perhaps the biggest gain of combining VISTools data with AI at sea is the possibility to build a digital twin of the entire North Sea. The first steps are underway, with detailed interactive maps of key fishing grounds. In the future, these maps will also incorporate data from new technologies such as eDNA. "With a digital twin of the North Sea, policymakers can simulate the impact of their decisions and move toward data-driven governance," says Hans Polet. "That makes it possible to protect vulnerable marine habitats and safeguard fish stocks, while still supporting fisheries and expanding the blue economy."

Digital Fields

Let's rewind to May 2020. European Commissioner Frans Timmermans has just presented his long-awaited food strategy for the EU: Farm to Fork. The strategy, part of the European Green Deal, sets ambitious targets for the agrifood sector. One striking element is the aim to cut the use of chemical crop protection products in half by 2030. Substantial funding is being made available, including for research into precision agriculture. This reflects the EU's twin ambitions for a future that is both *green* and *digital*.

Above and around ILVO's trial fields, drones and robots are already a familiar sight. On the roof of one of the campus buildings, drones rest silently in automated boxes, ready to fly out on their own and scan the fields. From behind their computers, scientists watch the live feed of images instantly processed by AI models. "We're testing whether a trained observer walking through a potato field is faster at spotting early signs of the fungus *Alternaria solani* than our drones," says ILVO scientist Jonathan Van Beek. "So far, it's a tie." *Alternaria solani* is a persistent challenge for potato growers. It causes economic losses through crop damage as well as the cost of chemical sprays. But it's tricky to detect the fungus early enough using sensors and AI. "Nuance is important in Europe's green and digital story," adds ILVO's Jürgen Vangeyte. "Research is advanced in some crops and for certain tasks. For example, distinguishing weeds from crops works well. Detecting individual Colorado beetles in the middle of a potato field works fine too. But in other crops and for other applications, we still have a lot of work to do."

The recipe for precision agriculture is clear, at least. By using drones and other sensors to monitor crops, farmers can respond more quickly and decide more accurately whether treatment is needed. They can then use precision hoes, burners, or lasers and as a last resort, a precision sprayer. By applying crop protection only where necessary, only the minimum of chemicals are used. The same principle applies to fertilization and irrigation. "This will allow us to produce food far more efficiently and sustainably. Instead of giving a field what we *think* it needs, we will *know* what it needs down to the level of each individual plant. Whether we will succeed in halving the use of chemicals across all crops and in every situation remains to be seen. But we will certainly come close. At the same time, we don't expect technology alone to provide the solution. We also focus intensively on regenerative, agroecological and organic farming systems, and we combine those insights with technological innovations. That's how we build integrated solutions that work for farmers *and* for the environment."

Smarter Application Techniques

An unusual lab can be found on the same campus where drones take off and land. Instead of microscopes and flasks, small drones and a frame fitted with small spray arms are passing over potted plants. Using paper strips and laser beams, researchers measure the exact spray pattern of the crop protection droplets as they land on the crop.

How much product lands on the plant and how much next to it? Is the spray uniform or patchy? "To achieve Europe's target of halving chemical crop protection, we also have to look at the application equipment itself. Products must be sprayed as precisely as possible, with minimal losses to the environment," explains ILVO scientist David Nuyttens.

In spray tech jargon, such losses are called drift. Starting in 2026, Flemish farmers will be required to use techniques that reduce drift by 90%, up from 75% in 2023. "In the spray lab, we test nozzles and new market-ready spraying systems, then assign them a reduction score. The most efficient systems today already use low-drift nozzles, trailing curtains, or air systems that guide droplets evenly and quickly into the crop. The most advanced machines combine all of these with sensors that detect weeds or diseases in real time, opening only the nozzle closest to the target."

Autonomous Robots

The next leap forward in precision agriculture will come from robots and so-called "active tools". The Test and Experimentation Facility (TEF), which ILVO operates on behalf of the EU, helps companies to prepare their robotic systems and implements for use in the field. It offers a "sandbox" environment for testing within regulations, as well as a test field and a workshop filled with technicians and engineers. In Moorslede, Belgium, the brothers Steven and Maarten Vanhoucke test what looks like a simple weeding blade. Their family business, Vanhoucke Machine Engineering, designs and builds agricultural innovations. Their blade is actually a smart tool that continuously exchanges data with the robot that carries it. "This means it only opens where weeds are present, so you can safely weed right up to and even between your plants," explains Steven Vanhoucke.

The robots' biggest advantage is how they address several problems at once. They are light, so they reduce soil compaction in comparison to heavy spray machinery. They work autonomously, eliminating the need for a driver and thus helping to address the farm labor shortage. They run on electricity, which means fewer greenhouse gas emissions. They are multifunctional, which makes them cost-effective. And they collect vast amounts of data close to the plant, improving the AI models that guide them. “Robots are the missing link that precision agriculture needs to reach its full potential,” says ILVO researcher Simon Cool.

What does this look like in practice? Farmers can already see it for themselves at the test field in Merelbeke-Melle. Back in June 2021, an audience watched as a drone flew over a maize field. Within half an hour, an AI model in the cloud had converted the drone images into a task map for a conventional sprayer. The contractor then drove across the field, but this time, the sprayer nozzles only opened above the weedy patches. The result: a nearly full tank after treating the entire field. That translates into 80% less product in comparison to a normal blanket spray.

Two years later, ILVO demonstrated the latest prototype of the Vanhouckes’ smart weeding blade. While it flawlessly removed the weeds, the robot carrying the blade also took photos to train a weed-recognition model. Minutes earlier, a prototype planter made by the Lauwers company had planted savoy cabbages while automatically uploading the location of each planting hole to the cloud. That data not only guided the weeding blade, but also allowed the AI model to train itself in a semi-automatic and energy-efficient manner. It received the instruction to treat everything in the photo as a weed except for the exact planting location. Just before this spectacle, the new Djust-e—a remarkably quiet, fully electric tractor—had prepared the soil without a driver in sight.

“Field demonstrations are crucial, because the success of these technologies ultimately depends on whether farmers adopt them,” stresses Jürgen Vangeyte. “They need to see with their own eyes how it works and be able to give feedback. Only then can we lower the threshold for them and truly improve usability in practice.”

Rules of Conduct and Climate Impact

Are farmers ready for this? Not quite yet. The systems are still often expensive, complex and not fully adapted for use in practice. But it is only a matter of time before broader data availability through data spaces and powerful multimodal AI models will increase both usability and relevance.

"At the same time, farmers must realize how crucial their own role is. The European data space and the accompanying regulations offer them a unique opportunity to keep control over their own data. They need to take an active role and extract value from it themselves. If they don't, other players—from technology companies and financial institutions to food companies and retailers—will step in and do it for them," warns Jürgen Vangeyte. The agrifood sector is becoming increasingly attractive to new entrants seeking predictability and added value, precisely because AI and data can help to better manage typical agricultural risks such as weather, disease or price fluctuations. "Data ownership is not a luxury; it is a prerequisite for future-oriented farming."

Are there any other risks? "It would be naïve to think otherwise. That is why Europe has taken the lead with its AI Act, which sets out clear rules of conduct. It distinguishes between applications with low and high risk for humans. Low-risk use, such as summarizing a report, comes with light requirements. High-risk use—such as assessing a subsidy file or an insurance case—requires an AI model to be transparent, explain its decisions and remain under human control. That is how we avoid problems with bias or faulty data," says Jürgen Vangeyte.

Even with such safeguards, precision farming still requires investment. And what about energy consumption? "That is a legitimate concern," Vangeyte acknowledges, "but I believe the balance will ultimately be positive. Yes, artificial intelligence requires computing power and thus energy, especially the large language models like ChatGPT, which run on billions of parameters in energy-intensive data centers. But in agriculture and fisheries we mainly use smaller, task-oriented models that run on batteries or on the device's own power supply. These are deliberately designed to be lightweight and efficient, because on a field, in a barn or at sea there is often limited access to electricity and internet. And don't forget the positive climate impact this technology offers. Precision farming means a reduction in use of raw materials, chemical crop protection, water and fertilizer as well as fewer food losses and lower greenhouse gas emissions. Ultimately we'll get higher-quality food for the 10 billion mouths the world will need to feed by 2050."

Precision Livestock Farming

Fast-forward to 2024. A group of farmers, journalists and technology developers gather at ILVO to watch as a camera scans trays of ground meat. No physical samples are taken, yet the visitors see information about freshness, fat content, moisture and origin appear on a screen. Are these samples of beef, pork, chicken, or lamb—or a mixture?

Down the road, the group sees a demo of a solar-powered mobile camera placed at the edge of a pasture. The audience watches as the camera follows each cow while artificial intelligence instantly interprets every movement as "standing," "lying," or "grazing." "That gives a farmer peace of mind, because you can monitor your herd remotely. The system detects restlessness, signs of illness or simply when it's time to move them to a more tender patch of grass. That matters not just for productivity, but also for animal health, climate impact and the biodiversity of the pasture itself," explains ILVO scientist Jarissa Maselyne.

Precision livestock farming uses data, sensors and artificial intelligence to support farmers in caring for their animals. The goals are to improve productivity, animal welfare and animal health while reducing the use of resources and veterinary drugs such as antibiotics. Sensors perform continuous measurements and perceive more than a human can. They can detect sooner when a piglet is sick, when a sow is in heat or when a cow is becoming lame. In some ILVO studies, researchers could even predict up to 24 hours in advance when a cow was going to calve. Such insights can help farmers avoid structural losses in the form of costs and yield losses. "Of course, the system must be tested, work well and be used correctly. The best way to think of it is as a good personal assistant, an extra pair of eyes and ears. Ultimately, it is still the farmer who decides what to do with the information," says Maselyne.

Ergonomics

Most systems also bring ergonomic benefits for farmers. Instead of driving 100 half-ton cows to the milking parlor twice a day and attaching each milking unit by hand, a barn with a milking robot handles the process automatically. These setups include advanced barn layouts that regulate cow traffic using chips located in neck collars. The gates open or close automatically based on the data in the chips. Has an individual cow has already been milked enough that day? Then the gate to the robot stays closed. If a cow shows unusual behavior—visiting the robot too often, or not often enough—the farmer receives a warning. The milking robot analyzes the milk on the spot and sends an alert when certain values deviate.

In pig and poultry farming, robots are already being used to automate routine tasks and issue warnings. Climate control and feed distribution in pig barns are already often automated. In poultry houses, robots aerate litter and prevent hens from laying eggs on the floor. "Right now, these are all separate systems. Every barn builder has something to offer, but the result is a jungle of apps and disconnected recommendations that are not easily usable for the farmer. Data sharing and AI models that integrate different sources into a single piece of advice will make these systems more practical," says Maselyne.

The Pig's Digital Twin

To measure is to know, but measuring alone does not always provide the solution, especially in the complex reality of animals reared in a barn. How do you prevent piglet mortality in the farrowing unit? How do you reduce antibiotic use? A digital twin may provide answers, just as it does in fisheries. "Instead of one parameter, a digital twin gives you a full picture of the herd. That picture also considers the environment and the relationships between animals. This makes it possible to predict growth potential and health for each individual animal, and to test on your own farm which interventions work best against stress or disease," explains Maselyne. This approach can also help farmers reduce the climate and environmental impact of their operations, while researchers can use it to avoid animal trials altogether. "For instance, with a digital twin, you can predict the impact of a new feed on nitrogen emissions just by running a simulation."

ILVO, KU Leuven and Ghent University are collaborating to develop a digital twin for pig farming. Meanwhile, they are working on AI models that already improve specific aspects of animal welfare. One model that predicts the behavior of sows and newborn piglets can be applied in different farm units to reduce piglet mortality. Another model is being trained to detect welfare issues based on cameras and sensors mounted in six participating European slaughterhouses. "It should provide feedback not only to slaughterhouses, but also to transport companies and farmers. This could improve animal welfare throughout the European food chain."

Whether precision livestock farming truly improves animal welfare remains a concern among welfare experts. False alarms or incorrect diagnoses could have negative effects on health and well-being. Adapting barn construction to maximize sensor effectiveness could disrupt natural animal behaviors. Technologies designed primarily for production efficiency tend to reach practice faster. Sensors may reduce direct contact between farmer and animal, leading to a sense of alienation between farmers and their animals, and a loss of farmers' unique stockmanship skills. "There is a risk that new generations of livestock farmers will know more about data and software than about animals, and that the 'instrumentalization' of farm animals will increase. Contact between farmer and animal may shift more to the negative side—intervening only when something is wrong—because those interactions are harder to automate," warns ILVO animal welfare specialist Frank Tuyttens. The future will determine the balance, but Maselyne is also aware of the risks. "A key point in developing the digital twin is to ensure that the farmer ultimately has more time to spend with the animals. The goal must be to bring humans and animals closer together, not further apart."

A Food Revolution

World Food Safety Day, 2024. A group of officials from Europe's national food agencies are visiting the Food Pilot in Merelbeke-Melle. They watch as a camera scans ordinary apples. To the human eye, they all look identical: four beautiful apples with a healthy red blush. But the AI model analyzing the images identifies the one most likely to rot first. Hidden beneath the skin, the camera has already detected the earliest signs of decay.

A few kilometers away, residents of the Lemberge Residential Care Center of Zorgband Leie & Schelde are being served their midday meal: leek soup, ham rolls with Belgian endive and mashed potatoes, with caramel flan for dessert. Before the plates are handed out, each one is placed on a scale with a camera. After the meal, the plates are scanned again to measure any leftovers. Based on the photos, an AI model estimates how much each resident ate for every meal, every day. The kitchen team can use this information to fine-tune menus, plan purchases more efficiently and cut down on food waste. The care staff also gains valuable insight into each resident's risk of malnutrition.

Back at the Food Pilot, food safety experts study a 2018 listeria outbreak. Comparative DNA research into the exact bacterial strain found in patients and in food products exposed the source. The technique used was whole-genome sequencing, which has quickly become indispensable in tracing and preventing foodborne infections. With this method, scientists can map the complete genome of bacteria and compare them at the strain level. Bacteria of the same species, such as *Listeria monocytogenes*, have highly similar genomes. But bacteria of the same strain, and thus the same source, have genomes that are identical down to the last base pair.

Bacteria and Their Fingerprint

"If a group of people suddenly gets food poisoning from *Salmonella* and at the same time a company reports *Salmonella* contamination in, say, a batch of eggs, these might be unrelated events. But if the *Salmonella* strain in the patients matches the one in that batch of eggs and you know the patients likely ate those eggs, then you can be nearly certain you've found the source of the contamination," explains ILVO expert Geertrui Rasschaert.

So far, 44,000 bacterial species have been described by science; most likely that's only a fraction of their real number. Fewer than 1% of these can make us sick, and only a handful are responsible for food poisoning. Still, their impact on human health can be enormous. According to the World Health Organization, 600 million people worldwide fall ill every year after ingesting contaminated raw meat or fish, uncooked vegetables, dairy, eggs and composite foods. In Flanders, with a population of some 7 million people, there were 302 major foodborne infections and food poisonings in 2023 that affected at least 1,332 people.

You have likely heard news reports of the most notorious foodborne bacteria: pathogenic *E. coli*, *Salmonella*, *Campylobacter* and *Listeria monocytogenes*. Noroviruses, parasites and molds can also cause problems. "Our labs already had highly advanced techniques to compare bacteria," notes ILVO director Lieve Herman. "But with whole-genome sequencing, each bacterium receives a unique fingerprint, making it possible to map out its entire transmission route."

Nanotechnology and biosensors will soon make this possible using portable, handheld devices, almost in real time. That means testing would no longer be confined to laboratories but could also take place directly within production facilities. Food companies would then be able to screen their products and environments for bacteria more frequently and at an earlier stage. "This can help them trace persistent microbes to their source, potentially saving countless lives worldwide."

Malnutrition

Back at the nursing home in Lemberge, the residents have finished their meal, and the kitchen is preparing the evening bread service. "Planning meals is a major challenge for residential care centers. Every day we prepare 200 meals for patients and residents with different needs, conditions and eating habits," says director Gert-Jan Andries. Across all hospitals and care homes in Flanders, an estimated 25 to 40% of food gets wasted. At the same time, half of all residents are at risk of malnutrition. Many have a high need for care, have difficulty swallowing or chewing due to illness or dental issues, suffer from trembling hands or have challenges related to taste and smell.

At the request of Gert-Jan Andries and several colleagues in the care sector, researchers are working on solutions. "These solutions lie in food preparation techniques," says ILVO scientist Geertrui Vlaemynck. She researches tailored nutrition for vulnerable groups and has collaborated with the organization Parki's Kookatelier to write specific cooking guides for Parkinson's patients. "By adjusting taste, presentation, composition and texture, as well as by playing with shape and color, we can provide safe, nutritious and appealing meals for people at risk of malnutrition or with problems chewing and swallowing. The psychological impact is significant, because eating is—and should always be—an enjoyable experience. For people who are very frail, we can enrich meals with proteins or other nutrients. Sometimes the only thing they can still eat is ice cream. In that case, we had better make sure that their ice cream is packed with proteins, fibers and antioxidants. In that way, even a small portion delivers the nutrition they need."

Malnutrition doesn't only lead to additional health issues. It can also undermine or block therapies for existing conditions. Wounds heal more slowly and medications become less effective. Good nutrition, on the other hand, can strengthen treatments or reduce symptoms. In Parkinson's patients, researchers from ILVO and KU Leuven found that eating certain fibers can enrich the gut microbiome and boost the production of short-chain fatty acids such as butyrate. This may help reduce tremors, which in turn makes eating easier again.

Systematic Screening

"The first step in preventing malnutrition is to systematically monitor eating behavior. In a care setting with many caregivers, such as a residential care center, that is not easy to do. Sensor technology and artificial intelligence can help," explains ILVO researcher Nathalie Bernaert. ILVO has built a prototype plate scanner that, with some adaptations, could be integrated into hospitals, residential care centers, schools and even company cafeterias. The aim is to continuously generate data and at the first signs of abnormal eating patterns, alert caregivers, just like the signals they already receive from medical equipment.

"A pilot project using the prototype has already taught us, at a collective level, how to cook in a tailored fashion and to make our purchasing more efficient. Which menus are popular? Which dishes consistently result in a lot of leftovers? What is the average consumption per resident? It also gave us feedback on our adjustments. If we prepare our vegetables differently, do people finish their meal or leave food on their plates? It's a tool that allows us to quickly optimize our processes," says Iris Lootens of Curando, the other residential care organization that partnered with ILVO to scan plate waste.

The "plate project" is receiving broad support across the care and catering sectors in Belgium and the Netherlands. Much of that success is due to NuHCaS, a knowledge network that brings together the fields of food, care and health. NuHCaS was founded in 2019 by the

Flemish knowledge and sector organizations ILVO, Flanders' FOOD, VIVES, POM West Flanders and TUA West. "Our mission is to improve people's quality of life and health by making better use of the potential of nutrition. One example is enabling better meals in care by fully leveraging available knowledge and technology. Automated data collection of eating behavior in hospitals and care homes is only the first step. We dream of a health data space modeled after the European agricultural data space, where data from many sources converge. This would allow us to not only better understand the links between health, food and care, but also to keep people healthier for longer," says ILVO scientist Marc De Loose, founder of the NuHCaS network.

DNA technology, artificial intelligence and data have the potential to spark a revolution for everyone working with food. "We can detect malnutrition earlier in vulnerable groups, trace the source of large-scale foodborne outbreaks more quickly, predict product shelf life more accurately and reduce food waste in a targeted way," adds Lieve Herman.

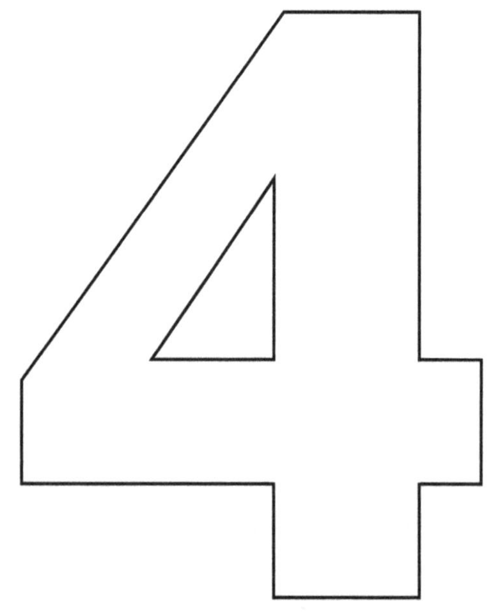

CHAPTER 4
Water and Land: How Do We Avoid Scarcity?

France, 2023. Six thousand people gather in the village of Sainte-Soline to protest the construction of a water reservoir. The village itself has only 400 residents, but soon it will host 16 reservoirs, together holding 6 million cubic meters of water. The developer is a collective of several hundred farmers. Multiple people are injured during the protest; explosives and tear gas are deployed. A year later, the scene repeats itself near La Rochelle. The rallying cry *"No bassaran"* echoes across the globe.

Ghent, 2016. Fernand Huts, owner of the logistics giant Katoen Natie, purchases 450 hectares of farmland located in the Netherlands from the Ghent Public Center for Social Welfare (PCSW) through a Dutch subsidiary. The land lies just over the border from Belgium, a stone's throw from the Ghent-Terneuzen port area. The price tag: €17.5 million. That's an affordable price per hectare, but because the land was bundled into one large lot, the tenant farmers were unable to compete. Ten years later, public outrage over the sale is still reverberating.

Water and land have always been sources of conflict in human history. Whoever controls them, controls food production. And the scarcer they become, the fiercer the struggle. The French protesters fear that massive reservoirs will deepen inequality between farmers in a changing climate. The outraged citizens and farmers in Ghent fear that agribusiness and big capital are pushing small farmers out of the land market. Are these isolated incidents—or canaries in the coal mine?

A Water-Scarce Region

In Belgium it rains about 190 days of the year, a little more than half the time. But it rarely rains all day. Converted into hours, it rains about 7% of the time. Whether that feels like a lot or a little depends on your point of view. During those wet hours, however, an annual average of 910 millimeters falls on Belgium, with a peak up to 1146.2 millimeters in 2024. "Even so, Flanders is classified as a region with a high risk of water scarcity," says ILVO water expert Sarah Garré. There are many reasons for this. "We rely heavily on rainfall to recharge our groundwater reserves since we lack large rivers that could serve as sources. At the same time, we need a lot of water for industry, agriculture and households. On top of that, we fail to make optimal use of the 910 millimeters of rain we do receive each year. Large-scale groundwater models show that less than 30% of that rain actually replenishes our reserves. The rest evaporates or runs off to the sea because of our many paved areas and our extensive network of drainage ditches. Climate change exacerbates this vulnerability. We can expect longer droughts, interrupted by intense periods of rainfall that we cannot capture, and more evaporation due to rising temperatures."

The designation as a water-scarce region already affects daily life, as many noticed during the chain of record-breaking drought years since 2017. This goes beyond bans on washing cars or filling swimming pools: our very food security is at stake. The common crops grown by Flemish farmers—potatoes, sugar beets, maize, wheat—need water most urgently while growing in spring and early summer. And that is exactly the period when we can expect less rainfall. VITO, a well-known Flemish institute for technological research, calculated that 40 to 50% of the potato harvest could be lost due to summer droughts. Demand for potatoes in Belgium, the Netherlands, France and Germany—the four largest potato producers in Europe—has not yet reached its ceiling. But structural shortages of quality seed potatoes, rising pest pressure and extreme weather conditions—both wet and dry—are putting yield security under strain. Is it coincidence that Agristo, a West Flemish family business and the world's largest exporter of frozen fries, is expanding its production capacity in India? The Indian climate and soils are well suited to growing potatoes for fries. And the cultivation expertise? That gets exported along with the potatoes.

What Is the Cost of Drought?

The problem of yield uncertainty is not limited to the Belgian potato sector. Agriculture across the Mediterranean basin—traditionally a major producer of thirsty crops such as tomatoes and zucchini, but also olives, fruit, wine, nuts and grains—is among the hardest hit. In 2022, drought wiped out a quarter of Spain's grain harvest and one-fifth of its fruit and forage crops. Across Europe, the annual cost of acute drought and structural water scarcity is estimated at €9 billion. If climate change drives the global average temperature up by 3°C, the frequency of droughts will double and annual losses could soar to €40 billion.

"Rainfall shortages often trigger a domino effect in farming regions. Irrigation is most needed when the water levels in rivers and groundwater reserves are dropping. Soils lose their capacity to pass on the little water they receive to the plants. Crops become more vulnerable to pests and diseases. This ends in a negative spiral of problems that reinforce each other and ultimately undermine our food security," explains ILVO water expert Sarah Garré.

It is in this context of rising uncertainty that French activists protest the construction of *mégabassines*—giant water reservoirs. Their demonstrations sparked existential questions both at home and abroad: Who has the greatest right to water? Do rights also come with responsibilities? And in times of shortage, whose water gets cut off first? It's a painful paradox of supply and demand: the moments where water is least available is when we all want it the most, from farmers and drinking-water utilities to industries and ecosystems. The expectation is that both drought and conflict over water will intensify. "Climate models cannot give us precise predictions of rainfall in the future," Garré notes, "but what is clear is that we are losing our temperate, predictable maritime climate. Extremes are becoming the new normal: we get both too little and too much water."

To prepare for this, Flanders launched the Blue Deal in 2020, a package of dozens of preventive measures. The Blue Deal generally follows expert advice: let as much water as possible infiltrate soils and landscapes, buffer it and reuse it. In short: make space for water, both literally and figuratively. At the landscape level, this means restoring natural river basins, expanding floodplains and desealing paved surfaces. On farms, it translates into building buffers and weirs, smarter drainage, minimizing waste and caring for the soil.

How Much Water Does Agriculture Use?

Although agriculture is only the fourth largest user of water in Flanders, farmers pay a high price for drought during the growing season. Between 2011 and 2019, the average annual agricultural water use was 56.4 million cubic meters. In the exceptionally dry year 2022, that figure jumped up to an estimated 74.6 million cubic meters. Usage was also well above normal in the dry years from 2017 to 2020.
Most of the water farmers use is pumped from groundwater. In normal years, drawing surface water from rivers and streams remains limited, but demand rises quickly during droughts. In 2022, nearly 9% of total agricultural water use came from surface water, despite pumping bans that were already in place at certain times of the year.

Water Consumption by Sector in Flanders (2022)

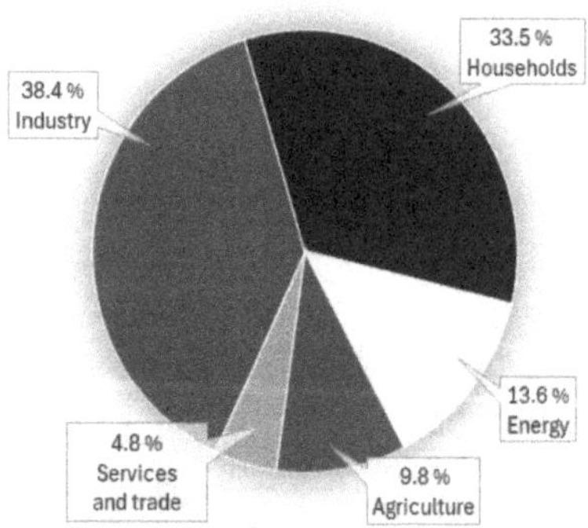

Figure 12: Total use of tap water, groundwater, surface water, rainwater and other sources (from products, ice, wastewater from another company or (drinking) water traded between companies). Cooling water is excluded. Source: Flemish Environment Agency (VMM), 2025.

Behind these general figures lie striking regional and sectorial differences. In West Flanders, for instance, agriculture is by far the largest water user (40%). Why? That province has a high concentration of livestock and vegetable farms. Across all provinces, arable farmers consume the least (1%), while greenhouse growers consume the most

(20%). Still, greenhouse horticulture is relatively sustainable in its water use: on average, 70% of the water comes from rooftop collection, and reuse is common practice.

Pig and dairy farms are the second- and third-largest consumers (19 and 14%), but their water use is less sustainable. "It's fairly complex," explains ILVO researcher Karen Goossens, "because they need water for a wide variety of tasks, each requiring a different level of water quality. This ranges from drinking water for animals to rinsing the milking equipment and cleaning the barn floors." Together with several partners, she is investigating which water sources are suitable for which purposes, so that livestock farms can also take steps toward reuse and conservation.

Farmers who grow vegetables or potatoes in open fields depend heavily on rainfall and soil moisture. When drought strikes at the wrong moment or lasts too long, they turn to groundwater or surface water to irrigate their crops.

Water Sources

Different water sources vary in quality and in their suitability for different applications. Tap water is potable and considered the highest quality. Because it is expensive, farmers only use it for specific, high-value uses such as rinsing milking equipment. Surface water pumped from rivers, streams and ditches is commonly used for irrigating vegetable and potato fields. Groundwater, drawn from deep or shallow aquifers via boreholes, is used for barn cleaning or as drinking water for pigs and poultry. Farmers need permits for this water source, which are increasingly difficult to obtain. Rainwater is either collected from rooftops, greenhouses or paved surfaces, or provided by wastewater utilities. Much less common is the use of treated wastewater or effluent from sewage treatment or industrial facilities; this water source is tightly regulated. Farmers may also purify their own water for reuse. Depending on its quality, this water can be used for irrigation, cleaning or cooling.

Irrigation

In Flanders, irrigation is mainly applied to high-value crops such as vegetables, potatoes and ornamentals. During a severe drought, irrigation can save a harvest and in some cases even protect crops from pests. A study on leeks, for instance, found that irrigation was just as important in preventing damage from thrips (insects) as chemical crop protection. But it also comes at a cost, both financially and in terms of labor—often at night.

"Irrigation doesn't have the best reputation among the public, but it is essential for some crops as drought becomes more frequent," says Sarah Garré. There is still much room for improvement in Flanders. The mobile reel irrigator often seen in fields today sprays water in a wide arc over crops. It applies a large amount of water in a short time, but distribution is uneven and the wind can easily carry droplets away. A boom irrigator, which has multiple nozzles to apply water more gently and evenly, is more efficient, but it's also more expensive and harder to move between fields. A fixed drip irrigation system that applies water drop by drop to the soil surface—or even just below it in the root zone—can use up to 30% less water than a reel. However, it requires substantial investment in installation and maintenance and cannot easily be applied to all crops.

Occasionally, these investments bring unexpected and sometimes impressive risks. "When farmers invest in costly drip systems, they understandably want to use them as much as possible and cannot afford to risk yield losses. In some cases, this actually leads to *more* water being used, especially when you zoom out to the landscape scale," explains Garré. She argues for combining irrigation technologies with sensors and crop growth models that predict exactly how much water each plant requires. "The sensors already exist, and the crop models are advancing quickly. Soon, we'll be able to fine-tune irrigation to the precise needs of every single plant," adds Tom De Swaef, ILVO crop modeling expert.

Crop growth models combine real-time cultivation data, satellite images, weather records and soil characteristics to predict crop development. They can forecast with great precision how much additional water a plant needs in order to thrive. But a model is only as good as the data on which it is built. That is why ILVO also continues to invest in fundamental knowledge about stress and growth in plants.

"Just like people, plants can experience stress from their environment, and they have different strategies to cope with it," explains Tom De Swaef. "In times of drought, they may root deeper, reduce transpiration by closing their stomata or curl their leaves. But in nature, these strategies usually come at the cost of lower yields. Think of desert plants—they remain dormant until it rains, then reproduce at lightning speed. That's perfect for surviving in extreme environments, but not for producing food."

Plants and Water Management

All plants need water to regulate their temperature, to grow and to feed themselves. They absorb water through their roots, but 90% of this water is lost again through evaporation via the stomata in their leaves. You can compare it to sweating—the evaporation prevents plants from overheating. Water also drives plant growth by literally putting the plant under pressure (turgor). This pressure in the cells arises when enough water enters through the roots and exits via the leaves. When plants wilt, that pressure is gone: growth stops, and the leaves and stems collapse. Finally, the internal water transport system distributes the nutrients taken up through the roots and leaves throughout the plant.

Soy, Quinoa and Chickpea

Around the world, scientists are searching for crop species that can tolerate drought while still achieving high yields. This is why three crops in particular have attracted so much attention in Belgium: soy, a subtropical crop; quinoa, which comes from the Andes mountains; and chickpea, a dryland crop. "All three are rich in protein, which fits well with the European strategy to reduce our dependence on imported protein," explains Hilde Muylle, ILVO expert on novel crops. "Quinoa also tolerates saline conditions, and both soy and chickpea are legumes, so they benefit soil fertility. Put all those advantages together, and you can see why people are investing in breeding programs. Obstacles do remain, however, and importing soybean meal from South America is still more profitable than growing soy in Flanders."

Hemp?

One striking newcomer among alternative crops is industrial hemp. It is grown for its fibers, seeds, oil and stalks—useful for textiles, materials, food and feed. In just two years, the area of hemp cultivation in Flanders has tripled. Even so, with only 512 hectares in 2025, it remains a very modest crop. Industrial hemp has no hallucinogenic effect; its THC content is too low. Farmers must apply for permission in advance and be registered as farmers or growers. Hemp is valuable for pollinators and can help remediate soils contaminated with substances like PFAS. Its deep roots make it fairly resistant to prolonged drought.

Does this mean that the "potato country" of Belgium will soon replace potatoes with these "exotic" crops? It's tempting to make a comparison to the rapid adoption of maize in Northern Europe during the 1960s and 70s, but such a large-scale shift is unlikely. Much depends on the world market and the geopolitical relations between exporting and importing countries. "The big difference with maize," notes Muylle, "is that there was no need to compete on the world market, nor did it need processors or traders. Farmers grew it for their own livestock and stored it in their own silos. Soy, quinoa and chickpea, on the other hand, are primarily intended for human consumption and sold to food companies operating in the global market. That means sharp prices and razor-thin margins. Belgium can compete in potatoes, but for soy, quinoa and chickpea, we still have a long way to go."

Better Varieties

That explains the investment in new varieties of existing crops, with the aim of combining drought tolerance with good yields under Belgian climate conditions. "And we're making good progress," says Isabel Roldán-Ruiz, scientific director of ILVO's Plant Sciences department. "Ever since the beginning of agriculture, people have been crossing plants to make them stronger. That's known as breeding."

ILVO has breeding programs for chickpea and quinoa, as well as for grasses, clover, plantain, cover crops and ornamentals. The institute also spent 10 years developing soybean varieties adapted to Northern Europe's relatively short growing season. That program was eventually spun off into the company Protealis, which now markets six successful soybean varieties. The entire soybean breeding process took about 15 years from the first parental selections to the approval of the first commercial varieties.

"Breeding is always slow. You start with 10,000 individual plants to eventually end up with one new variety. We focus not only on yield and drought tolerance, but also—depending on the species—on disease resistance, quality or cold tolerance. Drought tolerance is already complex in itself, because what good is a grass that survives a heat wave if it fails to recover when it starts raining again in autumn? That is why fundamental research on plant stress responses in interaction with soil and the environment is so essential."

In 2023, ILVO opened HYDRAS, a unique and strategic field lab for this type of research, with international attention and support from Flanders and the EU. Along the E40 highway, which runs from Bruges to Brussels, you can easily spot the large plastic domes or "rain-out shelters" near Merelbeke-Melle. Nowhere else in Europe can scientists monitor individual plant responses to drought or other stress factors such as salinity at such a large scale and with such accuracy, over an entire growing season, both above and below ground. Above ground, measurements are made with drones and specialized cameras, and below ground, a network of electrodes passes electrical currents through the soil to measure conductivity. ILVO encourages other research institutes and companies wishing to compare candidate varieties to present a project for HYDRAS.

"Many people are working hard to develop plants that can feed us in a changing climate. We expect to see significant progress in the coming decade," says Roldán-Ruiz. Hilde Muylle also stresses the potential and the necessity of new crops and varieties. "Yes, there is drought and extreme rainfall. But there are also soil fertility issues and the need for strategic food autonomy. These are all reasons to diversify our fields. The tipping point at which farmers and their buyers will follow is closer than we think."

Hairy Roots

New scientific insights sometimes emerge by accident. Researchers at ILVO and Ghent University took chrysanthemum leaves and fused them with a bacterium in a test tube. They hoped this would keep the plants compact without the use of chemicals, a study requested by ornamental plant breeders. The resulting plants did indeed remain compact (mission accomplished!) but they also proved to be more resilient to drought. A likely part of the explanation lies in the so-called "hairy roots" that developed after fusion with the bacterium. This serendipitous finding will need more study before it can be turned into a practical tool in the fight against climate change.

Soil Salinization

Back to the field and the crops being grown today. Potatoes and vegetables such as cauliflower and celeriac are highly sensitive to drought and thus require irrigation. But not just any water will do. Long-term irrigation with unsuitable water can damage the soil, as Israel, Egypt, Algeria and Cyprus have discovered. It is now common practice there to irrigate with treated municipal or industrial wastewater, which globally is used on about 10% of all irrigated cropland and grassland. In Flanders and the Netherlands, this is not yet common, though several pilot projects are being followed with interest. "Experiments confirm the potential," says Sarah Garré, "but they also show that the quality of the water must be carefully controlled before use." Regulations are very strict, requiring detailed monitoring of the substances present in the water to prevent irreversible soil contamination.

What's the problem? Wastewater often contains elevated salt levels, which over time can lead to soil salinization. Salinization is harmful both to crops and to the soil itself. Over the long term, it can render land permanently unsuitable for food production. According to the UN Food and Agriculture Organization (FAO), salinization is already one of the leading causes of soil degradation in Europe and North Africa. Today it affects more than 800 million hectares of farmland—around 9% of the Earth's land surface. Climate scenarios suggest the problem will intensify as rising sea levels elevate the level of saline groundwater tables, in combination with hotter, drier summers. Salinization can also occur when too much freshwater is pumped up, causing deeper, saltier layers to rise into the root zone. Experiments with quinoa, tomatoes and spinach in Morocco, Egypt, Italy and Belgium show wide variation in crop tolerance, but never a yield benefit. "Prevention is absolutely the only sensible strategy against salinization," stresses Sarah Garré.

Another challenge with treated wastewater is that it often has a high organic-matter content. Scientists investigated whether this could be an advantage, for instance by boosting carbon levels in poor soils, but instead they found the opposite: organic particles can clog soil pores, blocking water infiltration. In the worst cases, soils even developed a water-repellent layer.

Water Reservoirs

The simplest way to store water locally for later use is in reservoirs. This is already common in greenhouse horticulture, where rainwater is collected from rooftops, but where no roofs are available, groundwater or surface water can be stored instead. That was the plan in Sainte-Soline, France, where a collective of about 100 farmers wanted to pump water in winter, store it in reservoirs and use it for irrigation in summer, when pumping is prohibited. A judge revoked their permit, however, ruling that this would increase overall annual groundwater extraction.

In Ardooie, West Flanders, vegetable processor Ardo took a different approach. In 2019, the company built a 150,000-cubic-meter reservoir to store its own treated wash water. Fifty local vegetable growers, organized in a cooperative, can use it to irrigate 500 hectares of farmland during the summer. In Wallonia, the southern part of Belgium, the company was even more ambitious: through its subsidiary Hesbaye Frost, it acquired a 40-hectare site in Geer with old reservoirs that were drying up. The water levels have since been restored, and an irrigation network has been laid out to supply 1,000 hectares of farmland. The total storage capacity there is 550,000 cubic meters.

"Reservoirs for rainwater or treated process water are valuable, provided they are not scattered randomly across the landscape," explains Garré. "Ardooie and Geer are strong examples of shared storage and efficient irrigation infrastructure. The costs and benefits are largely shared among a group of local suppliers and users." Ardooie also generates solar power from panels installed over the basin, while in Geer the rewetting of the site has helped protect sensitive natural areas. Small individual reservoirs are far less efficient, however. "They are costly to build, and their capacity is always limited. No farm can store enough water to survive three months of drought. But as part of a broader strategy to reduce water shortage risks, they certainly have a role," Garré adds.

Even the 150,000-cubic-meter basin in Ardooie ran dry by September of 2022, an exceptionally dry year. In the wetter years before and after, however, it remained nearly full. This contrast illustrates that even cooperative reservoirs have their limits. Each spring, the member farmers must decide whether to subscribe to a fixed volume of water for the season and pay a deposit whether they ultimately use it or not.

Controlled Drainage

Storing water in the soil is often a better solution than storing it above ground in reservoirs. Large-scale groundwater models in Flanders show that every year, 10 to 30% of all rainfall is lost through drainage into ditches and streams. "Drainage has made farming possible on low-lying, wet soils, and it has served Flanders well. But today we know that year-round drainage is not always necessary and that we would do better to keep as much water as possible in the soil," explains Sarah Garré.

One solution for flat areas with well-drained soils is controlled drainage with adjustable weirs. This allows farmers to manage the water table in their fields as if they were putting a stopper in the bathtub. In both spring and late summer, when fields need to be sown or harvested, the "stopper" can be removed to lower the water table. After the machinery work is done, they put it back in. The water table rises again, naturally fed by rain or groundwater.

"Controlled drainage combines the benefits of conventional drainage—that is, soils dry enough to work without damaging them—with those of undrained soils, which have better and longer water storage for crops during the growing season. It holds water when it can and releases it when needed." Pilot projects show that this method can help postpone summer water shortages and, for high-value crops like potatoes or vegetables, even save one or two irrigation rounds on sandy soils. "That may not sound like much, but in our short growing seasons, having water for a few weeks longer can make a real difference. For maize, it can make or break the entire growing season."

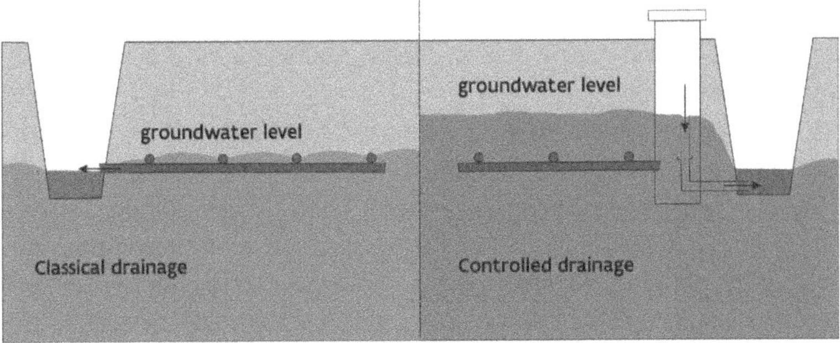

Figure 13: With conventional drainage, groundwater flows directly into a ditch as soon as it rises above the depth of the pipes. With controlled drainage, the water first enters a control well, where the farmer can insert or remove a regulating tube to set the water level. Without the tube, the system allows the water to flow directly into the ditch. With the tube in place, the water is held back until it reaches the top of the tube. ©ILVO

While conventional drainage has a poor reputation for climate adaptation, controlled drainage turns that reputation on its head. The good news is that much of the existing drainage infrastructure in Flanders can be retrofitted. The actual benefit for farmers depends on the soil type and the amount of groundwater naturally pushing up from below. "It requires an investment, but where conditions are right, it pays off—especially when farmers in a region work together to coordinate their water levels. In some cases, it can even boost yields, but regardless, it always retains more rainfall in the soil, which is an important societal benefit." Adoption is still slow and is mostly limited to the flat landscape with light sandy soils in the provinces of northern Limburg and Antwerp, where the system works best. Researchers are now also exploring its potential on heavier clay soils.

Nature-Based Solutions

In 2021, a so-called "water bomb" devastated the Belgian provinces of Liège, Namur and Luxembourg, killing 39 people and leaving many homes uninhabitable. After years of concern about drought, the disaster was a harsh reminder that we must also prepare for the possibility of flooding. Experts have a prescription ready: restore green-blue networks in open landscapes by reconnecting old riverbeds and expanding floodplains; increase surface roughness in sloping farmland by installing swales, hedges, weirs and tree lines; reduce paved surfaces in gardens, driveways, villages and cities; and manage wet-valley areas better.

"These are essentially the same measures that help against drought," explains Jeroen De Waegemaeker, climate adaptation expert at ILVO. "Reservoirs, sustainable irrigation and controlled drainage all have their uses, but they only become truly effective when integrated into a broader, structural landscape plan." Measures such as swales and wet grasslands can meaningfully increase the buffering capacity of the landscape. They also bring added benefits, from boosting biodiversity to cooling the nearby area. "But they require a broad, landscape-scale approach, best developed in consultation with all land users. Poorly placed swales—built without forethought or in the wrong spot—can cost a lot and solve little, and they can even create new problems."

Swales?

Swales are shallow ditches with accompanying berms constructed along contour lines to slow runoff and prevent erosion. They are often combined with shrubs or trees planted on the berms, which creates a more stable structure and potentially provides extra farm income in the form of timber or fruit. In the area around Voeren, for example, seven experimental swales—both temporary and permanent—have been installed in the Grenzeloos Bocageland Landscape Park, stretching a total of 765 meters. "Swales can be a valuable tool to capture, slow, redistribute and store runoff water. They can even recharge groundwater and prevent crops at the top of slopes from drying out while the ones at the bottom get waterlogged. But they are not suitable everywhere. Their irregular shape takes up space, and their effectiveness varies depending on soil type and slope. Farmers need to think carefully about whether the effort and land use are worthwhile in a specific situation," says ILVO scientist Bert Reubens.

Figure 14: Working principle of a swale. Runoff is collected in a shallow ditch with a planted berm behind it, allowing water to infiltrate the soil. This way, soil particles and nutrients present in the water are also retained rather than lost. ©ILVO

Regional Coalitions

Jeroen De Waegemaeker and his colleagues have studied various regional projects involving farmers. "Projects where farmers join together to buffer water or manage groundwater levels tend to be more successful and they last longer than individual efforts such as investing in a private reservoir. A single measure on a single field is not necessarily the most efficient solution, even for the individual farmer," says ILVO researcher Sylvie Fosselle.

This principle led the Flemish government to launch the Water-Land-Schap (Water Landscape) program in 2017. Farmers, companies, residents and landscape managers get support to take joint action on water. Some 30 projects are already underway. One example is the project called "Climate-robust valley of the Molenbeek/Gondebeek and Driesbeek". The East Flemish provincial government works together with farmers to install weirs and restore a historical streambed, while the municipalities of Merelbeke-Melle and Oosterzeke are taking action to address water quality and erosion. "This water project is part of the greater Rodeland landscape project, where 19 project partners are working together to create a valuable natural and agricultural landscape. Everyone gains something from it, and that is tremendously inspiring," says Rodeland coordinator Sofie Van Brussel.

Rewetting

Sometimes landscape interventions are so drastic that coalition-building becomes difficult and conflict can arise. The city of Geel lies in eastern Flanders, in the valley of the Kleine Nete and Aa rivers. Sixty years ago, a prestigious and unprecedented development project drained the marshlands there. Fourteen dairy farmers were carefully selected and resettled behind protective dikes in the valley, as a symbol of humankind's triumph over water. Farmers from all over Europe came by bus to drive across the dikes and admire this showcase of modern agriculture.

Today, the atmosphere is very different. At a meeting with the provincial governor, the current farm managers—sons, daughters and grandchildren of those carefully chosen pioneers—listen to the Flemish Government's plans to rewet the very land on which their future was built. Many words are spoken about nature, river flows, greenhouse gases and evolving insights. "Paludiculture" and "new business models" are presented to the group. But these families find it a bitter pill to swallow. You cannot keep dairy cows in a swamp. So what is a dairy farmer supposed to do?

"It is easy to say they should just switch to paludiculture or wet crops. But a first assessment based on experiences abroad suggests that farming this way in Flanders will not keep a farm going," says Jeroen De Waegemaeker. Paludiculture refers to wet crops such as cattail, reed, miscanthus, giant duckweed, sphagnum moss and cranberry. Technically, these crops show promise, but Flanders still lacks the markets, adapted varieties, cultivation know-how and machinery. Cattail straw, for example, is suitable for mushroom cultivation, but the intensive drying process makes it too expensive as a raw material. It can also serve as a peat substitute in potting soil or as a green input for board and insulation materials, but there is no logistical chain in place to collect the small, scattered harvests and deliver them to the processors. The Netherlands, Germany and Ireland tell a different story: they still have large peatlands and the scale of wet crops is large enough to build viable supply chains.

"This is the familiar bottleneck we see with every new crop," says Jasmine Versyck of the Flemish B2BE Facilitator, which stands for "business to bio-economy." "We help the farmers experimenting with innovative crops to connect with potential buyers. These are not just the classic food companies, but also producers of building materials, cosmetics, pharmaceuticals, textiles, paints and more." The emerging bio-economy could hold an opportunity for farmers, provided Europe is serious about building a green, fossil-free economy. "It could help to diversify cropping systems and provide an additional income stream for farmers who face rewetting or who need to broaden their farm operation for other reasons. But it is pioneering work. You have to learn a new language—do you speak agriculture or do you speak chemistry?—and that is not for everyone," Versyck concludes.

Feasible Farming

To reconcile farming with rewetting, Jeroen De Waegemaeker points to five interlocking "gears" of research and innovation. The first gear is to have clear environmental goals that look far enough ahead and do not "change horses in midstream"—a frequent criticism during farmer protests in the Netherlands and Flanders in 2024. The second is to explore a broad range of crops and systems, from small-scale rewetting in conventional crops or grassland to extensive rewetting using wet crops. "For example, to what extent is it still possible to keep cattle in wet areas, given certain adjustments and different breeds? How does pest pressure increase on wet plots and which crops can still be cultivated?" The third gear is realistic farm business models, which can only take shape if the fourth and fifth are also present: creating a market with a logistical chain and providing additional support—financial or otherwise. "We must work on all of these gears at the same time to offer a promising future to farmers like those in Geel," says De Waegemaeker.

And why does this matter? "You have to realize that farmers are the ones maintaining our landscapes. They are the ones who plant hedgerows, create buffer strips and sow field edges. We talk about making our landscapes more resilient to the impacts of climate change, but that has a direct effect on their income, while at the same time we are asking them to make an extra effort. They bear the heaviest burdens and the greatest risks. Farming must stay feasible if we want to keep the countryside alive."

Water Quality

There is no one-size-fits-all solution for reconciling agriculture and water. There are indeed many urgent questions to ask, such as about the economic value of water. About ownership, with its rights and obligations, and about water quality, which is becoming an even greater challenge in a warming climate. The health of Flanders' rivers and streams has improved since the 1990s but still falls far short of the legally required standards. By 2027, Flanders aims to have at least 15 of its 194 waterways classified as clean. As of 2025, only one waterway makes the list. Water quality is a complex issue, and the sources of pollution are diffuse. Roughly 300,000 households in Flanders are still not connected to the sewage network. Their wastewater flows untreated into streams and rivers. This pollutes smaller watercourses, according to the 2025 citizen-science project called "Watermonsters" (or "Water Samples" in English), run by the newspaper *De Standaard*, Waterland vzw and KU Leuven. Another concern is the limited capacity of our sewage system and the 9,000 overflow points that discharge excess wastewater into streams and rivers during heavy rainfall to relieve pressure on the system. Add to that the places where industry is still permitted to discharge wastewater into rivers. Especially in rural areas, agriculture contributes to peak loads: fertilized soils leach nitrates and phosphates, and crop protection products leave residues in the environment. Too-high concentrations of nitrates (nitrogen) and phosphates (phosphorus) lead to eutrophication of waterways. The most visible effect of this is excessive algal growth, which harms other plants and fish.

Water Quality in Flanders

The Flemish Environment Agency (VMM) monitors water quality in Flemish waterways through a network of measuring stations and modeling. Samples are taken of surface water, groundwater, fish and smaller aquatic organisms, and information is gathered on emissions of phosphorus, nitrogen and oxygen-consuming substances from households, industry and agriculture. Only 7% of larger water bodies score well for phosphorus, and 36% for nitrogen. Phosphorus, a pollutant mainly emitted by households, has slowly improved since 2007. Nitrogen, mainly from agriculture, stagnated during the dry years from 2017 to 2020. “We know that dry years are usually bad years for water quality. Crops take up less nitrogen from fertilizer, so there is a higher risk of leaching. With the same application rate—within legal limits—the risk of nitrate residues in the soil is greater during a dry year. That is difficult to anticipate,” says ILVO expert Thijs Vanden Nest.

Residues of chemical crop protection products are generally found less often in streams and rivers, although a small number of products continue to cause problems. In 2025, 1,2,4-triazole made the Belgian news when it was found in excessive concentrations in drinking water in West Flanders. The fauna and flora are faring better, although the majority of waterways remain in moderate to poor condition. A limited number of harmful substances such as mercury, PBDEs (flame retardants) and PFOS have been found to exceed environmental standards ranging from “sometimes” to “often”. The origins of these pollutants are not always clear.

Climate change is expected to make it even harder to clean up waterways. Drier summers mean that pollution is less diluted, while more extreme rainfall means more wastewater ending up in streams. “Agriculture, industry, sewer managers, water treatment companies and households all face the same challenge. It is important that we tackle it together. Technology can help us to apply fertilizers and crop protection products more efficiently. Another part of the solution lies beneath our feet: a healthy soil that is less prone to erosion does not leak nutrients into the environment and retains water longer. Soil restoration measures are also water measures,” says Sarah Garré.

One such measure is mulching. A trial at the HYDRAS field lab showed that celery mulched with a layer of grass-clover retained water longer during drought and was also better able to use the water from a sudden rainstorm. “Mulch helps keep moisture in the soil. That reduces evaporation losses and increases water availability for the crop. It also improves soil structure, which allows water to infiltrate better and reduces erosion,” says ILVO researcher Maarten De Boever.

Still, mulch is no miracle cure. Beyond the question of whether enough materials are available, mulch can sometimes hinder crop growth. That was not the case in the celery trial—quite the opposite—but not all biomass is equally suitable for use as mulch. “Success depends heavily on the type of mulch, the soil you are working with, the weather conditions and the crop. It is not always useful, but in the places where it works and where materials are easily accessible, mulch can be beneficial.”

Speculating on Land

Back to Ghent, where the Ghent PCSW lands were sold off in a single large parcel to businessman Fernand Huts. Does he intend to take up farming when he retires? Perhaps. But more likely, he is seeking a safe investment, betting on rising land values or building a land reserve to use later in land swaps for port expansion. Speculation on farmland is nothing new. In both Flanders and Wallonia, farmland prices continue to rise well beyond agricultural value. What is new is the large-scale public outcry. Eight years after the initial transaction, the Huts affair drew more than 15,000 visitors to the 2024 exhibition “Gentse Gronden” (“Ghent Farmlands” in English) at the Ghent City Museum (STAM). The exhibition was based on doctoral research by Hans Vandermaelen (ILVO-Ghent University) and his co-curator Esther Beeckaert (Ghent University). Vandermaelen’s research was the first to map the scale and evolution of public ownership of farmland in East Flanders.

Figure 15: Announcement of the exhibition about the public lands owned by the Ghent Public Center for Social Welfare (PSCW) at the Ghent City Museum in 2024. ©STAM

Ten years earlier, Vandermaelen’s colleague Anna Verhoeve had carried out equally groundbreaking doctoral research on the actual reuse of farmsteads and farmland after sale. She found that, on average, 15% of agricultural land is not used by professional farmers. These are parcels officially zoned for agriculture, but in practice they are used as gardens (6.2%), as pasture for backyard animals (4.3%), or for other non-agricultural purposes. At the same time, Flanders loses 4.6 hectares of open space every day to housing, industry, recreation, infrastructure and more. “Agriculture is under pressure in rural areas, and that leads to more urbanization. Not only the loss of access to land or farmsteads is troubling; the number of complaints and challenges to permit applications is rising. In short, this limits people’s commercial freedom,” says Anna Verhoeve (ILVO).

Several factors drive these developments. One is the flexible spatial planning policy that allows luxury residences and non-agricultural businesses in agricultural zones. That’s why you see houses with five-hectare gardens, landscaping firms, construction companies, daycare centers, riding schools, restaurants and even wellness centers located on farmland and in former farmsteads. There is also the macroeconomic logic of capital and the fact that land is considered a safe investment. Sometimes speculation plays a role, or land is purchased to be exchanged later. “Whether private individuals, landscapers or industrialists who are doing the speculating, they all bring financial power that far exceeds the agricultural value of the land and buildings. This drives up land prices. Bit by bit, fertile soil is disappearing, and along with it the basis of every viable farm.”

For farmers, access to land is the literal foundation of their farm operation. Land access generates income and autonomy, it is a prerequisite for permits and it determines how many animals you may keep, how much manure you may spread and much more. For society as a whole, farmland also has value: it preserves open space, produces food, provides cooling on hot days, buffers carbon and water and shapes the landscape. Should we be concerned that farmers are losing access to land? Yes, because it is happening quickly, and it is usually irreversible. In 2024, the average age of Belgian farmers is 56, and only 13% of those over 50 say they have a successor. In the coming years, many farmsteads and parcels of land will come onto the market. “That could be an opportunity, provided there is an active land and property policy. Leaving everything to the free market will only accelerate the loss of agricultural space.”

Public Lands

It would be a mistake to narrow the discussion about farmland to one of rising land prices alone. That is indeed a real and growing problem, but farmers do not necessarily have to own the land they cultivate. This is not a revolutionary idea: leasing farmland has given farmers security for generations and still does. In addition to cultivating the land of private owners, for centuries farmers have also cultivated land owned by governments or semi-public institutions such as church parish councils. In total, governments in Flanders and Brussels own more than 214,000 hectares of land—17% of the total parcel area. Almost one-third of that (31%) is cultivated by farmers.

How Much Land Is Publicly Owned in Flanders and Brussels?

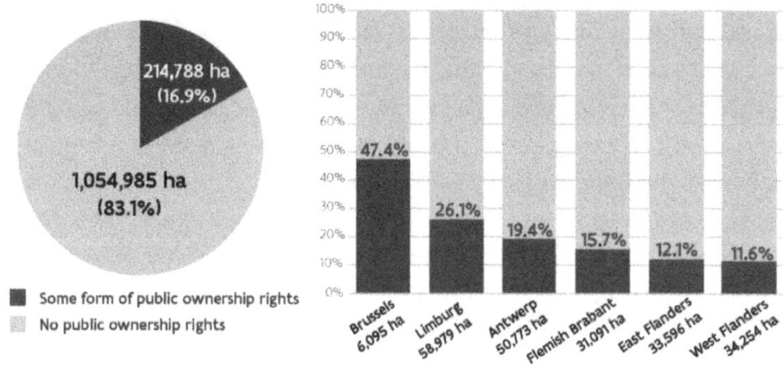

Figure 16: Share of public land ownership in Flanders and Brussels relative to the total parcel area. Public roads, rivers, or squares are not included. On the right, the same data for Flanders, broken down by province. Source: ILVO, based on cadaster data (FPS Finance), 2024.

"Much more than ownership, what truly matters for farmers is long-term security of land use. Those who could lease land from a PCSW or a parish council had no worries, because public authorities were known never to sell their land and, unlike private owners, they do not die. But that historic reality is fading, since in recent decades PCSWs and parish councils have begun actively selling their lands," says Hans Vandermaelen.

Together with his colleague Glenn Willems, he found that after many centuries of stability, the number of sales by PCSWs and parish councils has risen since the early 2000s, and not by a little: 30%

(PCSWs) and 10% (parish councils). This stands in sharp contrast to the active purchasing policies of other public authorities, who buy land for entirely different policy objectives. "Over the past 20 years, governments have been fairly active on the land market, with a structural increase in total public land ownership as a result (+10%). But those acquisitions have served goals related to nature, water, mobility, or industry. Seldom, if ever, were they made in support of agricultural policy," says ILVO researcher Glenn Willems.

The recent wave of PCSW and parish land sales directly affects thousands of farms currently working those lands. "For these farms, the looming threat of an imminent sale is a major and sometimes paralyzing source of uncertainty. When land actually does come onto the market, farmers often lose out. The zoning designation may remain 'agricultural,' but the actual use often shifts. And those farmers who do manage to purchase the land are saddled with heavy loans that may compromise the transfer of the farm to the next generation. In most cases, this means a loss for agriculture," says Hans Vandermaelen.

A Public Lever?

The reasons for these sales are well known. Local authorities need money for other investments, and lease income is relatively low. On top of that, PCSWs often own farmland outside their own territory. The Ghent PCSW's holdings over the border in the Netherlands are a well-known example. The PCSWs of Bruges, Kortrijk, Brussels and Mechelen also own large amounts of farmland well beyond their administrative borders. "Although this was never seen as a problem historically, public institutions today argue that it is not easy to manage such lands as policy instruments. They focus on potential market value and neglect the use value. It is crucial to bring that use value back into view, if only to make sound policy choices."

That use value includes agriculture, landscape management, open space and carbon storage. But it can go further still. Long-term access to public land can give farmers the breathing room to experiment with new, sustainable practices. Investing in soil health or regenerative farming is far easier on land you can count on using for years to come. It is much harder on land held under expensive, short-term leases that must give an immediate return on investment. “You cannot expect farmers to make long-term investments in their soil if they aren’t sure if they will be able to keep farming it. Public authorities hold an enormous lever with their farmland to offer that certainty, and in doing so, they can deliver on policy goals beyond agriculture.”

An experiment in the Grenspark Groot Saeftinghe, on the left bank of the River Scheldt, shows that it can work. The European Union wants to restore the population of the marsh harrier, a bird of prey that hunts small mammals in grasslands and wetlands but is struggling to survive. Fifteen farmers there have been granted long-term access to public farmland by the Flemish government in exchange for adopting “marsh-harrier-friendly” agricultural practices. “It is a fine example of a win-win for agriculture and nature. The government achieves multiple objectives through an active land policy. Moreover, it creates a snowball effect extending beyond the boundaries of public land, since these farmers are now also applying the new crops on their private land.”

An Active Policy Is Essential

The urgency is clear. Agriculture is under pressure in rural areas, and current attempts to protect it are failing. The total area zoned as agricultural is not shrinking, but in practice, more and more parcels are being used as gardens, hobby pastures or for other economic activities. Public authorities hold a trump card with their farmland, but it is not being played. “There is fear of active policy-making, and even the agricultural sector itself is wary,” says Elke Rogge, scientific director of ILVO’s Social Sciences department. “‘Where will government interference stop?’ they ask. ‘Freedom of cropping is absolute and must be protected.’ But the result is counterproductive. The absence of an active farmland policy is fueling unwanted dynamics. To anchor farming in the Flemish countryside, we must take an area-based approach.” Elke Rogge’s department published a blueprint for such an approach in a widely disseminated policy advisory note (2024). Below is a summary. This is what needs to happen, dear reader. This is what we expect of you, dear governments, landowners, mediators and real-estate agents.

Observatory

A first step is the creation of an observatory for the Flemish agricultural land market that monitors ownership, transactions and prices. Is there a quiet consolidation underway? Is there foreign involvement? Accurate data is the foundation for a serious debate about farmland. For inspiration, we can look to Wallonia's *Observatoire du foncier agricole*.

Land Bank

Establishing a Flemish land bank with a clear agricultural program and structural financing would enable an active farmland policy. This is the only way to prevent the land of farmers who stop farming—more than 4,000 between 2025 and 2029—from getting fragmented or effectively disappearing from agricultural use. The Flemish Land Agency (VLM) already manages several land banks and occasionally exchanges farmland, but this occurs largely as an ancillary tool to support other domains such as nature, industry or infrastructure.

Non-Agricultural Developments

The creeping conversion of farmland into gardens or hobby pastures must be curbed, as it is putting increasing pressure on agricultural activity in Flanders. A stricter approach is needed to address non-agricultural reuse of farmland and farmsteads. Negative recommendations or filing appeals against permits are not enough. The Flemish Government's decree on non-conforming land-use changes must itself be revised, and additional (financial) conditions and obligations could be imposed on permit requests.

Pre-Emption and Demolition Fund

The creation of policy tools and incentives could help redevelop agricultural sites for farming purposes. Is a site suitable for reuse? Then a system of pre-emption rights or guided land mobility could facilitate takeovers by young or beginning farmers. If not, then de-sealing and demolition are good ways to prevent further fragmentation of farmland and farm buildings in the countryside. A demolition fund could compensate the loss of value for the farmer selling the land. The fund could be financed by the levies imposed on non-agricultural land-use changes.

What about heritage value? Only 7% of the remaining farmsteads are deemed valuable enough to keep.

Public-Land Policy

The sale of public land must, at the very least, be called to a (temporary) halt so the consequences can be fully assessed and the debate can be held in earnest. Governments, both Flemish and local, can play an important role in reconciling conflicting interests between sectors and administrations. They can act in an advisory capacity to help establish appropriate land-use agreements (lease, long lease, temporary transfer of use).

Area-Based Approach

An area-based agricultural vision is needed, with clear goals and explicit choices. Each part of the Flemish countryside has different development opportunities: which functions are desirable where? Structural cooperation between the Agriculture and Environment policy areas is crucial here, including the involvement of local authorities and other stakeholders active in the terrain. Existing area-based programs such as Water-Land-Schap and the landscape parks can serve as a starting point.

Area-Based Workers

Permanent area-based workers could become the driving force behind this geographic approach. They know both the policy framework and the local cases, actors and contexts. They could serve as the point of contact for farmers and conduct bilateral preparatory discussions. For inspiration, we can look to Milan (*distretto agricolo*) or other highly urbanized regions in Europe.

Dear reader, will farming be all right? Will our food supply be all right? It will be, if we take this type of advice seriously.

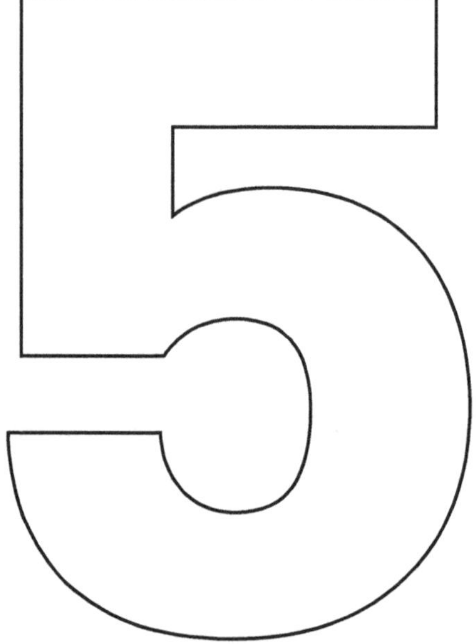

CHAPTER 5

New Gene Technology: the Hen that Lays Golden Eggs?

Picture this. You're reading a book about food. You've been at it for a while. You just finished a chapter on water (it made you thirsty), and now you arrive at a chapter with chickens and eggs in the title.

Your stomach growls. You open your fridge—or your backpack if you're on the train—but you hesitate. The book is calling you back; it's all so terribly fascinating. Luckily, you still have a banana. Perfect for a quick bite.

Then your smartphone pings. The news app headline reads: "The end of the banana." In the first line you read: "The only banana variety found on our store shelves is doomed." You nearly choke on your coffee.

Wait, didn't you hear something about coffee a while ago too?

Indeed. Banana and coffee are symbols of the risks we face from genetic narrowing in extreme monocultures. The banana's popularity comes from its convenience: easy to eat, easy to carry anywhere. In Belgium, it's always number one or two in fruit sales, sometimes swapping places with the apple. But unlike the apple, the banana has no long tradition of breeding and improvement. Most bananas we eat come from genetically identical plants. Because cultivated bananas produce no seeds, farmers can only propagate them by cuttings. As a result, more than half the plantations worldwide are clones of a single mother plant: the Cavendish.

The danger becomes clear if we look back only a few decades. In the mid-20th century, the Gros Michel banana filled grocery store shelves. Like the Cavendish, it was a fine dessert banana. It was tasty and had a thick skin that made it easy to export. The Gros Michel was a big success, until a fungal disease swept through plantations in the 1950s. All plants were genetically identical and thus equally vulnerable to the fungus. The solution was the Cavendish, as it was resistant to that strain. But now a new variant of the same fungus has emerged, one the Cavendish cannot withstand, and history threatens to repeat itself.

Coffee tells a similar story of a genetic bottleneck, though less extreme. Two species dominate global production: Arabica, prized for its flavor, and Robusta, valued for yield and disease resistance. Yet genetic diversity within these species is also very limited, and Arabica—the flavor driver—is especially sensitive to climate change. Chances are you've already noticed: between 2022 and 2025, supermarket coffee prices rose by nearly one-third due to poor harvests.

Agrobiodiversity

The lack of genetic diversity in bananas and coffee is extreme, but according to the UN Food and Agriculture Organization (FAO), it affects all food crops to some degree. There are more than 6,000 edible plant species, yet 66% of our diet comes from just nine of them: rice, maize, wheat, potato, soybean, sugarcane, sugar beet, cassava and palm oil. This is an unintended side effect of highly successful, science-driven plant breeding that took off between 1940 and 1960. Farmers gained access to high-yielding varieties of wheat, maize and rice that outperformed the old, local varieties by 20 to 30%. These breakthroughs saved millions of people from starvation in South America and Asia. One of the pioneers, American Norman Borlaug, was awarded the 1970 Nobel Peace Prize for this achievement.

But as the global free market expanded, it also created dependency on a narrow range of commercial crops and varieties. The development of so-called F1 hybrids for maize, rice and tomatoes accelerated this trend. F1 hybrids are offspring from two pure parent lines with superior qualities, a biological effect known as heterosis or "hybrid vigor." Any harmful recessive genes from one parent are masked by stronger dominant gene versions from the other. The result: F1 hybrids have fewer weaknesses and the combined strengths of both parents. These hybrids are also genetically uniform, a clear advantage in large-scale farming. All the plants grow and fruit in the same way, at the same time.

F1 hybrids remain central for crops like maize, rice, tomatoes and many vegetables. They offer high quality, but they are expensive to develop. Farmers cannot save their own seed because the next generation of plants loses its hybrid vigor. Growers must buy fresh F1 seed every year from a shrinking number of seed companies. Only a handful of firms currently control half the global seed market and thus the gene pool of our most important crops.

Why Plant Breeding?

"Modern plant breeding has given us much to be grateful for. We are still reaping those benefits. But the loss of agrobiodiversity, combined with the growing monopoly of a few firms over the genetic diversity of crops, poses risks we cannot ignore," says ILVO director Johan Van Huylenbroeck. "Genetic diversity is essential for food security. It is our first and strongest line of defense against crop failure. The greater the diversity of species and varieties, the more resilient plant production becomes, and the less chance that pests or diseases will wipe out an entire harvest. But diversity is also the foundation for breeding itself: the more variation we have, the more new variations we can create."

Plant breeding is the process in which humans deliberately select and cross plants with desirable traits, hoping to combine those traits in the offspring. The traits of interest shift over time. High yield will always matter, but today disease resistance, drought tolerance, nitrogen efficiency and quality all rank equally high on the list. Breeding and selection also apply to livestock, where priorities have also evolved over time. The Holstein-Friesian dairy cow was the result of more than a century of strict selection with black-and-white bulls that produced cows with very high milk yields. Today, feed efficiency, health, fertility and even the shape of the udder and amount of methane emissions are part of the equation.

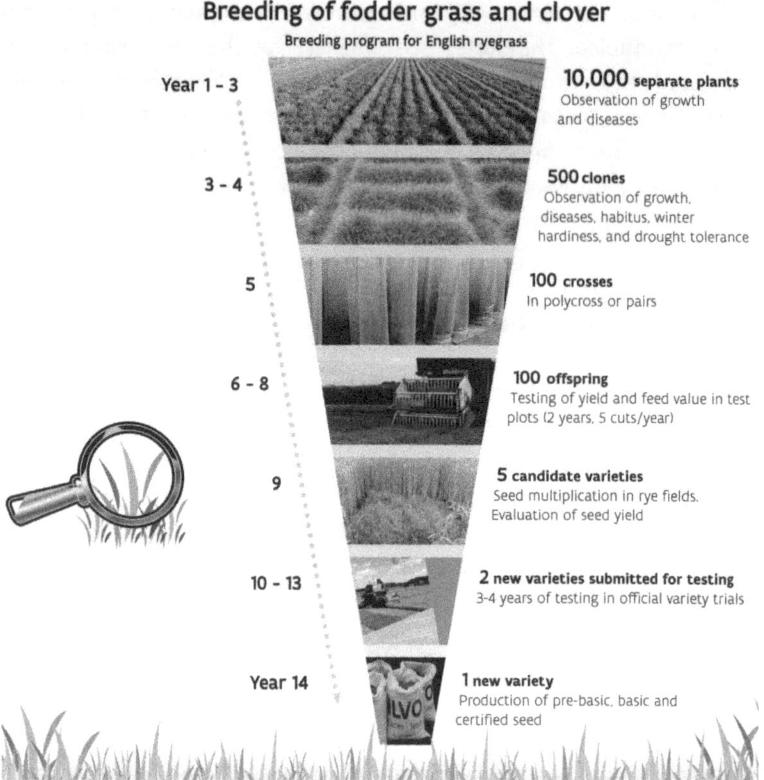

Figure 17: Timeline of a current ILVO breeding program. ©ILVO

In breeding, the first step is always to search for individuals with promising traits. The more genetic variation you have to start from, the better. A few decades ago, it was standard practice for breeders to collect plants from the wild, plant them in test fields, and evaluate traits they thought farmers and growers might need 10 years down the road. In this way, they expanded the gene pool available for breeding. Today, the tools are different—genes are stored in digitally searchable gene banks—but the same principle remains. What follows is a long process of crossing, selection, multiplication and above all, lots of testing. On average, it takes 10 to 15 years from the first cross to the registration of a new variety.

Crossing?

Just like humans, most plants have male (pollen) cells and female egg cells that must come together for fertilization to occur. Their reproductive organs are located in the flowers: the anthers contain pollen cells and the pistil contains an ovary filled with egg cells. To cross plants, the breeder applies pollen from the father plant to the pistil of the mother plant. This can be done using a small brush or by letting pollinators do the work inside a closed tent. The fertilized flowers then produce fruits containing seeds, which the breeder sows. The resulting seedlings hopefully carry the best traits of both the mother and father plant. This process of sexual reproduction is used for crops like tomatoes, maize and wheat. Sometimes breeders use other methods to create "mutations," or changes in individual plant cells. In that case, no seeds, fathers, or mothers are involved; instead, they apply ultraviolet light or chemical agents. The treated cells are grown in a petri dish. The rest of the process is the same as for seedlings created by crossbreeding: strict evaluation and selection.

Quality Control

Breeders cannot simply release new varieties onto the European market. They must first prove that the new variety can be clearly distinguished from others and that it consistently produces uniform plants with stable traits. For agricultural crops, there is also a trial in which candidate varieties are tested against established reference varieties. Only if they perform as well or better are they admitted to the European market.

"This system of quality control has ensured that today's varieties are an improvement over the older ones," says ILVO researcher Joke Pannecoucque. She coordinates the variety trials conducted in Flanders for the official Belgian variety lists. From an exceptionally large trial she supervised, it became clear that new varieties of Italian ryegrass—the typical silage grass stored for winter feed—produce up to 20% higher yields than varieties from 20 years ago. That means dairy farmers today need 20% less land to feed their cows with silage grass for an entire winter. For wheat, breeders successfully focused on varieties with shorter stems, higher grain yields and improved protein content for baking, coupled with resistance to fungal infections. In maize, breeding has delivered earlier ripening kernels with a higher starch content. In both crops, the genetic gain averages about 1% yield increase per year, despite the major leaps made back in the 1960s. In potatoes, disease resistance is key, but so is ease of processing into fries. Tomatoes and strawberries have gained better flavor and shelf life thanks to breeding, while rice varieties now contain lower levels of cadmium and arsenic.

Less headline-grabbing, but highly important for sustainable farming, are the advances in cover crops. Before 1990, there was not a single variety of mustard or fodder radish cover crops resistant to the sugar beet cyst nematode, whereas today every variety on Germany's leading recommended list can reduce nematode populations in the soil by over 70%. The same applies to root-knot nematodes. "By sowing these resistant cover crops before or after a main crop, farmers can significantly reduce infection pressure on their fields without having to rely on chemical crop protection," explains Tim Vleugels. This represents a considerable environmental and societal benefit.

Public and Private Breeders

Tim Vleugels is a plant breeder at ILVO. Together with his colleagues, he develops new generations of cover crops, as well as local varieties of chickpea and quinoa. The research institute also runs a renowned breeding program for grasses and other forage crops. It was given this explicit mandate in 1932 at the founding of the then-Federal Plant Breeding Station. The initial aim of the breeding station was to support the emerging livestock sector in Flanders. "Our initial goal was to match the famous 'fat meadows' of the polders in the Bachten de Kupe area with equal grass quality elsewhere in Flanders," recalled former Administrator-General Erik Van Bockstaele at ILVO's 90th anniversary in 2022. That goal was quickly achieved with the successful forage grass called "Vigor". Since then, over 100 ILVO-bred grasses have made it onto European recommended lists, and in 2025 two varieties even received top rankings.

"We choose deliberately not to breed maize or wheat, because the large private breeders already invest heavily in those crops. Our added value as a public breeder lies elsewhere: in smaller or niche crops that currently have less commercial value but are important for agro-biodiversity and sustainable farming," says Kristiaan Van Laecke, head of ILVO's Plant Sciences department.

Since the rise of modern breeding in the 1960s, the seed market has become highly concentrated. Today, just a handful of companies control half of the global market. Bayer Crop Science (Germany) invests heavily in maize, soy, cotton and rapeseed; Corteva Agriscience (U.S.) has a strong position in maize, soy and sunflower; and DLF (Denmark) dominates in grasses. Also prominent are Syngenta (Swiss-Chinese), Limagrain (France) and KWS (Germany). "In the past, every country had a public breeder. That is no longer the case. But besides the big players, there are still many smaller, specialized breeders." Belgium counts two public breeding institutes: ILVO in Flanders and its French-speaking equivalent CRA-W in Wallonia. ILVO specializes in forage crops, while CRA-W focuses on potatoes, cereals and fruit. France has the leading INRAE institute, which breeds grains and peas, while the Netherlands has Wageningen University (WUR), with its spin-offs that create innovative crops for the bioeconomy.

Belgian Hops

Hop may spark the imagination, but this crop has nearly disappeared in Belgium. Now it's being revived through the development of region-specific aromatic varieties. Despite Belgium's strong tradition as a beer country, most brewers and hop growers still rely on old or foreign varieties. Those often come from the United States, France, or the Czech Republic and do not necessarily match the Belgian "terroir" or soils. At the request of brewers seeking distinctive flavors for their specialty beers and after a successful call to the Belgian public in 2020 to collect wild hop plants, ILVO established a plant collection that now forms the basis for breeding. The entire collection was screened for potential, and the first successful crossbreeds are already growing in the fields of Poperinge. You can expect a new range of specialty beers brewed with Belgian hops, though they will first need to undergo several more years of testing before they reach your glass.

New Breeding Techniques

Back to your banana. We mentioned that banana plants are propagated by cuttings in plantations. This is because they are sterile, which also means they cannot be improved through conventional crossbreeding. Considering the importance of bananas as a staple food for more than 400 million people worldwide and the growing pressure on Cavendish bananas caused by a fungal disease, this is a serious limitation. Banana breeding is so difficult that it is rarely attempted—except in Leuven, Belgium. KU Leuven is a world leader in banana breeding. Its living collection of more than 1,500 varieties makes it the largest banana gene pool in the world. The university is now using that genetic wealth to pioneer a method that could make Cavendish bananas resistant. The method is based on CRISPR-Cas, a new breeding technology that promises to be a game changer in crop improvement.

Since the 1960s, we have quickly changed the crops and the varieties we eat, and the techniques used by breeders have also evolved. For centuries, breeders searched for diversity in nature or in the field, but later they began creating variation—"mutations"—themselves. The first generation of scientific breeders developed techniques using UV light and chemical agents. They also devised ways to bypass natural crossing barriers, for example by altering the number of chromosome sets in plants or by extracting embryos from ovaries and growing them in test tubes. This broad set of "classical breeding techniques" remains important today and continues to be refined.

Mutations?

Every gene in a plant's DNA encodes a specific trait and has a particular function. While the function itself cannot change, the expression of the gene (how strongly or weakly it is expressed in the plant) can. In the case of a growth-related gene, a mutation of that gene could cause a plant to grow faster or to remain more compact. Such a mutation can occur when the plant's DNA is damaged at that specific site. During the repair process of the damaged DNA, errors can occur, and these errors are called mutations.

Mutations occur constantly in nature because plant DNA is continually exposed to environmental factors. Cauliflower, Brussels sprouts and broccoli are examples of vegetables that arose through spontaneous mutations in a common ancestor. The same process happens in animals (and humans). The well-known Belgian Blue cattle breed, for instance, originated from a spontaneous mutation that led to extreme muscle growth. But mutations can also be induced intentionally, using classical or modern breeding methods. These methods differ widely in their level of precision.

A second major leap came in the 1980s when scientists Marc Van Montagu and Jozef Schell from Ghent University in Belgium discovered the principle of gene transfer from bacteria into plants. Their work laid the foundation for gene transfer technology, better known as genetic modification (GM). With genetic modification, you can isolate DNA from one plant and insert it into another, even across species that are not related and cannot be crossed using traditional methods. This was revolutionary, and the technique remains widely used today, although in Europe strict regulation of genetically modified organisms (GMOs) has largely confined it to medical and pharmaceutical research.

GMOs

The first genetically modified organisms released worldwide were soybean, maize, cotton and rapeseed varieties developed by large seed companies. These were engineered for insect resistance, but also for resistance to the companies' own herbicides. The latter, in particular, damaged the public image of GMOs, sparking mass protests and heated social debate. Around the world, trial fields were stormed and destroyed, including in Flanders. A trial with GMO potatoes in Wetteren, Belgium, conducted by Ghent University, ILVO, VIB and University College Ghent in cooperation with Wageningen University, was partly lost when activists ripped out a quarter of the crop. Large-scale protests and strict European GMO legislation brought most GMO development in Europe to a halt. In the meanwhile, a broad scientific consensus has emerged that GMOs are safe. They do not pose any unique risks.

Today, we are already far into the third great leap. A new generation of gene-editing techniques, known as new breeding techniques (NBTs) or genome editing, has emerged—including the CRISPR-Cas technology that KU Leuven aims to use to rescue the Cavendish banana. Like GM, these new methods directly alter a plant's DNA, but they do so more efficiently and less invasively. They harness the plant's natural DNA repair mechanism to create mutations at a specific, preselected location. In this way, they can induce variations that could also occur spontaneously in nature or be created with classical methods. Now those "natural" mutations can simply happen on demand, more quickly and more precisely.

CRISPR-Cas works as follows: a Cas protein acts as a molecular pair of scissors, making a precise cut in the DNA of a plant cell. An RNA "guide" marks the exact site. After the cut is made, the cell automatically repairs the break using its own natural repair machinery. During this repair, small errors arise, which are the mutations that breeders always try to create. You can even supply a short "template" of genetic code to make a highly targeted repair, such as including a piece of DNA from another crop that is known to confer disease resistance.

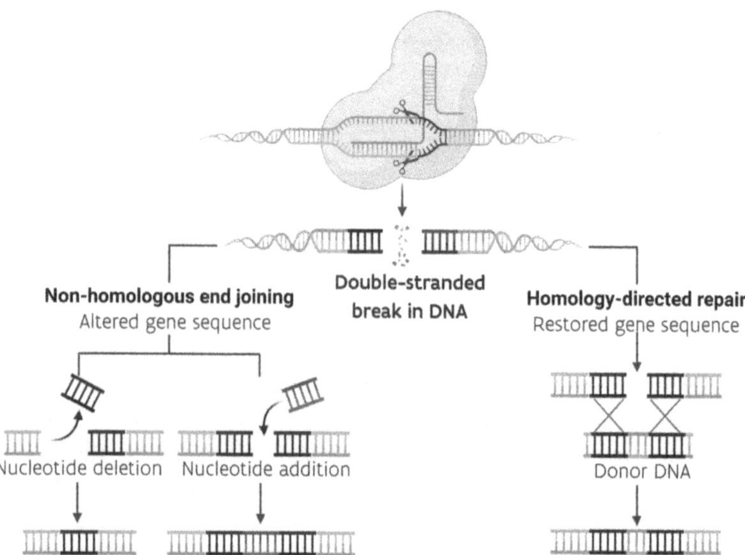

Figure 18: Inducing mutations in a plant's DNA with CRISPR-Cas. Created in Biorender. ©ILVO

Borrowed from Bacteria

Like genetic modification, CRISPR-Cas is based on a natural mechanism first discovered in bacteria. Bacteria defend themselves against viruses by adapting their own DNA. They copy and paste fragments of genetic code from viruses into their own genome so they can respond more quickly when faced with a new infection. This mechanism was first described in 1987 by Japanese researchers, although at the time they had not yet identified a name or function, let alone an application for it. In 2012, two researchers, Jennifer Doudna (USA) and Emmanuelle Charpentier (France), clarified the mechanism and its use in gene editing. Their work triggered a breakthrough, and by 2013 the first applications in mammals and plants were already underway. Barely two years later, the journal *Science* named CRISPR-Cas the scientific breakthrough of the year, and in 2020 the two scientists were awarded the Nobel Prize in Chemistry.

Fast, Precise and Accessible

CRISPR-Cas is just one of several new gene-editing techniques, but it is by far the most talked about. Unmatched in precision, efficiency, speed and versatility, it allows breeders to create more variation in the gene pool of choice, far more quickly. This can significantly shorten the entire breeding process. “It used to take months or even years to introduce a change into a plant’s DNA. With CRISPR-Cas, this can be done in just weeks or months. The risk of unwanted effects is lower because we can create a very precise, small mutation at a specific location in the DNA where we already understand the impact,” explains ILVO expert Katrijn Van Laere.

The potential for making agriculture more sustainable is considerable, not least because CRISPR-Cas is simpler and more affordable than genetic modification. “Affordable” means a faster track to profitability, even for highly specific applications with limited commercial value. These new gene techniques promise faster progress in disease resistance and drought tolerance in major crops. They can also provide a major boost for niche crops in smaller agricultural regions. Researchers are already working on more nutritious sorghum for subsistence farmers in Africa and low-cost, pest-resistant eggplants for farmers in Bangladesh.

"Bitter Genes" in Chicory

Scientists from ILVO, Ghent University and VIB have succeeded in identifying and switching off the genes responsible for bitterness in chicory using CRISPR-Cas. This demonstrates the power of the technology, but why is this useful? Chicory is processed into inulin, a healthy, water-soluble prebiotic dietary fiber. But chicory is naturally bitter, so manufacturers currently have to remove those bitter-tasting compounds using an energy-intensive chemical process. Non-bitter chicory roots could be dried and ground directly. "In addition, this research improves our understanding of how bitter compounds are produced in plants. With that knowledge, we can now screen large groups of plants—including vegetables such as endive, Brussels sprouts and arugula—for the presence and function of their bitter genes. That will allow us to develop not only less-bitter varieties, but also more-bitter ones, and everything in between," says Katrijn Van Laere (ILVO).

In parallel with these new biotechnologies, in recent years a host of other techniques have been developed which are equally important for breeding. Today, breeders have a large toolbox of instruments and methods to accelerate progress in the lab, the growth chamber and the field. Flying above ILVO's experimental fields and the HYDRAS field lab described in Chapter 4, drones collect vast amounts of data on crossbreeds and candidate varieties. That data, in combination with information from environmental sensors, underground moisture sensors, satellite images and mathematical models of crop growth, give breeders additional insight into how individual plants grow and respond to stress. "Breeders have a trained eye; they know what to look for when selecting plants. But with high-throughput phenotyping, they can collect and process massive amounts of data from large populations of plants to make even better decisions. This process will reduce the amount of time, resources and effort being wasted," says Isabel Roldán-Ruiz.

European Regulation

In Ghent, a spin-off of VIB is preparing to push the boundaries of breeding once again. Rainbow Crops aims to combine the power of the new gene-editing techniques with the advantages of artificial intelligence to improve multiple complex plant traits at once. It's a lofty promise: more efficient crop varieties that have good yields and are drought-tolerant, heat-resistant and disease-resistant at the same time. Globally, Ghent continues to play a leading role in biotechnology, even in the European context where legal frameworks for these technologies are slow to take shape. The rapid pace and potential of the science stand in stark contrast with the sluggish pace of regulation.

In Europe, crops obtained through genetic modification are subject to a heavy approval process: strict safety assessments, a vote in the European Council, transparent labeling and traceability requirements. For now, these same rules also apply to the new generation of gene-editing techniques. But Europe is now developing a revised legal framework that will distinguish between techniques that make small changes to DNA without gene transfer and those that bring about larger modifications. In the first case, which covers the vast majority of agricultural applications, there will be few restrictions beyond traceability. But traceability is precisely what is worrying scientists. How can you trace a gene-editing technique if it leaves no trace or marker in the final product? "A tomato with a CRISPR-induced mutation is molecularly identical to a tomato in which the same mutation occurred spontaneously," explains Marc De Loose. He heads the ILVO lab that specializes in detecting GMOs in the food chain. He is working with fellow researchers to determine how Europe can enforce the new rules efficiently and cost-effectively.

He understands the need for greater transparency about how, why and when these technologies are used. "It fits into Europe's overall vision of a transparent food chain, where citizens are informed about what they eat. That means we must develop systems and build infrastructures that allow us to achieve that goal in a practical and affordable way." He advocates for an integrated approach and sees opportunities in data technology. A first proof of concept of this approach was already successful: in 2010, Marc De Loose and his colleagues managed to detect and remove an illegal GMO from the market. The case involved a dietary supplement developed in Denmark, produced in Ukraine and commercialized in Poland—without authorization for sale or use in the European Union.

European plant breeders and biotechnologists are hopeful that the new regulations will soon make innovation easier in their field. The same is true for the precision fermentation discussed in Chapter 1. In other parts of the world, where these technologies are already allowed without restriction, crops are being developed that use 25% less water. Wheat can be planted that doesn't need any chemical crop protection, and there are apples that brown more slowly to reduce food waste. "New breeding techniques are not an end in themselves, nor are they a miraculous solution. But they are an essential part of the path toward sustainable, climate-resilient agriculture. We must be able to put all these puzzle pieces together to continue to provide everyone with enough safe and nutritious food," concludes Marc De Loose.

CHAPTER 6

Adventures in the Kitchen: Is Betty Crocker Ready for a Makeover?

Fourteen percent. That's the share of today's household budget that goes to food. Two hundred years ago, that figure was 60 to 80%! One hundred years ago it was still about half.

And what foods do we buy? If we were to believe all the media coverage of food hypes, it would be chia, matcha, kimchi, yuzu... Does this sound like you?

Chances are your shopping cart looks a bit more conservative. Perhaps it contains an avocado—a newcomer that has secured a permanent spot in the vegetable aisle—but most likely it is still filled with tomatoes, carrots and onions, the evergreen top three of fresh vegetables. And even though you might occasionally skip breakfast—the intermittent-fasting trend is clearly visible in the rising numbers of adults regularly skipping breakfast (11% fewer report eating breakfast since 2014)—you're still far more likely to stick with the rhythm of three meals and two snacks a day.

The Western lunchbox is still likely to contain a sandwich, while dinner is a home-cooked hot meal, more often with meat than fish or plant-based alternatives, and more often accompanied by potatoes than pasta, rice or another side dish.

Still, what we eat and how much we spend on it has changed spectacularly in a relatively short period. Where our ancestors in the 18th and early 19th century spent most of their modest income on a monotonous diet of rye bread, potatoes and buttermilk, today's supermarket shelves appear almost endless. For that, we can thank economic growth and industrialization, which have given us wealth as well as modern transport and storage methods.

Will the digital wave and the data revolution—which are still in their infancy—bring about another transformation of that magnitude?

Our Diet, Mirror of Society

In Leuven, birthplace of the Flemish Farmers' Union but also of brewing giant AB InBev, the Center for Agrarian History (CAG) studies the agricultural and culinary heritage of Flanders. Their research offers fascinating insights that link what we eat with broader societal shifts. "What, where and how we eat differently today than our (grand)parents says a great deal about the societal changes we have experienced since 1945," says historian and CAG coordinator Professor Yves Segers (KU Leuven).

Our plates now hold more meat, fried food and processed products than 75 years ago. We drink less milk and more soft drinks, wine and carbonated drinks. We consume more sugar and less bread and potatoes. New dishes are conquering our kitchens, often inspired by culinary traditions from other continents. We also eat out more frequently, in all sorts of ways. From a diet based largely on what we grew ourselves, we have shifted to one marked by convenience, diversity and market dependency. "The transformation was and still is gradual, but the result is a fundamentally different food system. After World War II, farmers, suppliers, food companies and retailers became increasingly interconnected. At the same time, through specialization and trade, the chain grew longer and longer. Power relations shifted, and those shifts have left their mark," says Segers.

Recent Agricultural History in Belgium

Up to 1850: Farming in Belgium is mostly small-scale arable agriculture for household use. Farm animals are used for traction and manure. The balance between food production and population growth is fragile, with crop failures often leading to famine. Sixty to eighty percent of the household budget goes to food. The daily menu: potatoes, bread and buttermilk, sometimes supplemented with eggs, cheese, meat or sugar.
1850–1950: Industrialization upends everything. Modern transport and the liberalization of the international grain market make long-distance trade possible. A flood of cheap imported grain from the U.S. and Ukraine, combined with industrial jobs as an alternative livelihood, force farmers to mechanize and shift toward livestock and horticulture. The 1920s bring prosperity and a more varied diet.
1950–1980: The end of WWII marks the end of the liberal trade policy. The following decades are defined by a managed agricultural economy and protected markets. Belgium focuses on specialized livestock farming and horticulture. By the 1960s–70s, the downsides of the growth model become apparent. Europe's most recent agricultural policy has serious drawbacks: environmental issues surface, animal welfare comes under fire, the number of farmers drops and they lose power in an ever-lengthening food chain. Affordable food, purchased in new supermarkets, becomes the norm.
Since 1990: Europe cautiously returns to a more liberal trade policy. Milk quotas, once a symbol of a managed economy, are abolished. Greening becomes a condition for farm subsidies. Alongside large export-oriented specialized farm operations sprout smaller, more locally oriented family farms. The middle class now spends only about 10% of its household budget on food.

Evolution in the Number of Farms in Belgium, 1980–2020

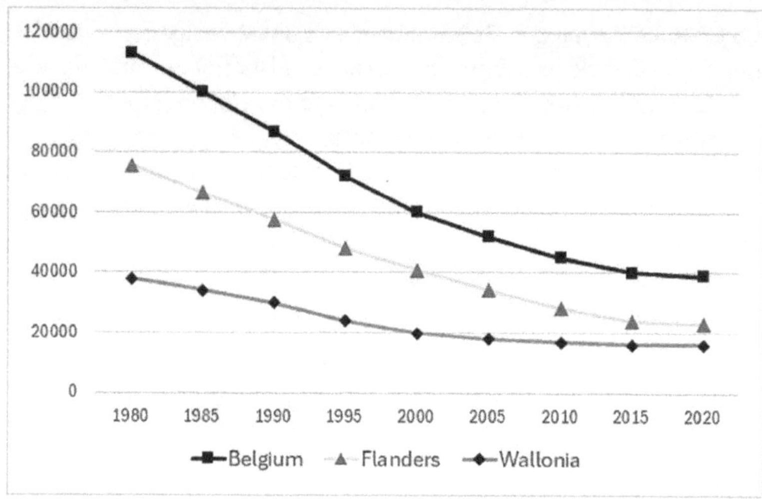

Figure 19: Flanders' Agency for Agriculture and Sea Fisheries–Agricultural figures: www.vlaanderen.be/landbouwcijfers and Statbel (Directorate of General Statistics–Statistics Belgium), 2025.

The Impact of Politics and Trade

The transformation had already begun in the late 1940s with the founding of the Benelux in 1944 and the European Economic Community (EEC) in 1958. The economic success that followed, together with the expansion of social security policies, ushered in the golden 1960s. Prosperity rose and so did the population, from 8.5 million Belgians in 1950 to 10.2 million in 2000. Agriculture had to adapt quickly. Mechanization, specialization and scaling up ensured success. Farmers shifted from producing cheap staple foods like grain and potatoes to products with more added value such as vegetables, fruit, meat and milk. Between 1960 and 2000, horticultural output rose by 212%, far outpacing the 85% growth in arable crops. Bell peppers and mushrooms appeared alongside new types of meat, dairy products and fruit varieties. At the same time, export markets grew exponentially. Pigs, poultry and vegetables were and still are shipped abroad in large volumes. Today, the tiny region of Flanders is the world's largest exporter of frozen vegetables and potato products. To sustain that growth, vast amounts of raw materials for feed and fertilizers were imported, making the chain from raw material to finished product longer in every sense.

Developments in the Food Industry

This rapid growth brought new challenges for the food industry. Clients and consumers became more demanding, and the distribution sector grew into a powerful negotiating partner. Food safety in production and storage became a guiding principle. The 1999 dioxin crisis in Belgium thrust food safety into public awareness. One year later, the Federal Agency for the Safety of the Food Chain (FASFC) was established. To turn the food industry into the hypermodern sector it is today, high-tech innovations were introduced: heating techniques to extend shelf life, liquid gases for ultra-rapid freezing and ingredient separation methods. Packaging also evolved quickly. Until the 1970s, paper, glass and tins dominated the market. Today, sterile and modified-atmosphere packaging are the norm.

The Dioxin Crisis

Despite the focus on food safety, things sometimes go wrong. In spring 1999, the Belgian dioxin crisis erupted. A sudden spike in poultry deaths and a sharp drop in egg production raised suspicions. Investigation revealed that both were caused by feed contaminated with PCBs and dioxins. These toxic compounds had entered the food chain through chicken and eggs. Egg-containing products were preemptively pulled from store shelves. Chickens and pigs were slaughtered en masse, exports of poultry, pork and beef were halted, and the production of cookies and Belgian chocolates was suspended. The crisis turned into a political scandal that cost several ministers their posts, not so much because of the health risks but because of attempts to cover up the danger to public health.

Many food companies could not afford the necessary investments on their own and merged, creating ever larger multinationals. Some remain headquartered in Belgium—most famously AB InBev in Leuven—while others merged with foreign players. Côte d'Or, the iconic Belgian chocolate maker, was taken over in 1987 by Jacobs-Suchard, which in 1990 was itself acquired by the American Kraft Foods/Mondelēz, becoming the world's third-largest food producer today. Antwerp's cookie producers De Beukelaer and Parein were also sold after a merger in the 1960s, first to the French group LU, then to the same American company that owns Côte d'Or.

The rise of the supermarket

At the same time, the distribution sector became a particularly strong player. The biggest change came with the arrival of supermarkets, a trend that began gingerly in the first half of the twentieth century but gained real momentum in the late 1950s, inspired by the American example. Delhaize opened its first self-service store with shopping carts in 1957 on Flagey Square in Ixelles. GB followed in 1958 with a store in Antwerp's Luchtbal district. Products on the shelf became ever more varied: fresh, dry, frozen, exotic or traditional—everything was available. Shopping became faster too, fitting the busier lives of dual-income households. Families drove there in the cars they could afford since the 1960s and stored their purchases in their refrigerator or freezer at home. Today, supermarkets remain dominant, accounting for 90% of all food purchases.

Household shopping behavior quickly became a major subject of study. Marketers and retailers used the findings to fine-tune packaging design, pricing, product image and even shelf placement. Consumers were divided into segments—teenagers, yuppies, young mothers, active 60-somethings—and targeted with a tailored marketing strategy. With loyalty cards and the rise of the internet, shopping behavior could be tracked digitally with even more ease. One outcome of this research, combined with the technological innovations implemented in the food industry, was that everything on the shelves started to look alike. Whether shopping at Delhaize, GB, or Colruyt, you could find the same products in the same packaging, placed in the same spot on the shelf. Supermarkets sought ways to differentiate themselves and found the answer in private labels. GB was the first to launch generic products in Belgium in 1978, again inspired by the American example. These were and still are a resounding success.

More Meat and Soft Drinks

The exceptional economic growth in Belgium between 1955 and 1973 also drove strong gains in purchasing power, averaging 3.6% per year. The impact on shopping, cooking and eating was immense. One of the most striking changes was the surge in meat consumption. Compared with 1948, a Belgian worker in 1978 ate twice as much pork, four times as much ham, five times as many cold cuts and an impressive 19 times as much chicken. In the past 60 years, worldwide meat consumption has nearly quadrupled.

Other foods that saw a marked rise in consumption included margarine, cheese, pasta, pastries, fruit and sweets. Drinking habits changed too. Milk lost popularity, while beer consumption first climbed, then began to decline in the 1980s. The biggest change, however, came with the arrival of soft drinks. By 1959, 70% of Belgian factories already had a vending machine for soft drinks. But the real breakthrough came in the 1970s, when more than three-quarters of secondary school students drank cola at lunchtime. To this day, Belgium ranks among the world's top cola-drinking countries, with consumption averaging 69 liters per person per year.

Evolution of Global Meat Consumption, 1961-2050

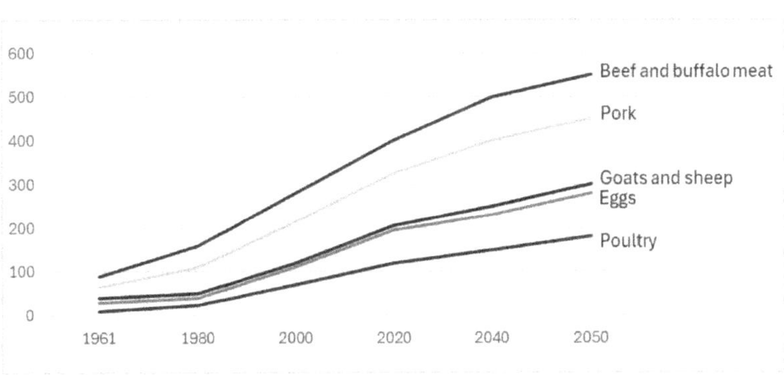

Figure 20: Expressed in million tons. Data from 1961 to 2013 are based on published FAO estimates; data from 2013 to 2050 are based on FAO projections, taking into account expected population growth as well as projected regional and national economic growth. Source: United Nations Food and Agriculture Organization (FAO), OurWorldinData.org/meat-production | CC BY, 2025.

Obesity

Today we eat more abundantly and more varied than in the past, but our diet is less healthy: we eat too much meat, fat and sugar, and too few vegetables, fruits and whole grains. On average, Belgians now consume 3,700 kilocalories per day, while our bodies need only an estimated 2,000 to 2,500. This overconsumption leads not only to a careless attitude toward food and food waste, but also to a growing obesity problem.

Trends in the Kitchen

From the 1960s and 1970s onward, more women started working outside the home. Cooking and shopping became a shared task in dual-income households. Time spent in the kitchen decreased, partly because there was less time after work, but also because of the help provided by appliances such as mixers, refrigerators, freezers, microwaves and dishwashers. Meals were generally less heavy than in people's childhood days, made with new, often exotic ingredients and frozen foods. Ready-to-heat meals and eating out also made their entrance. Around 1960, the average Belgian household spent 1.6% of its budget on restaurant visits. Today, that figure fluctuates between 5% and 6%, a spectacular increase.

In the 1990s, *nouvelle cuisine* peaked with so-called molecular gastronomy. Even at home, cooks began experimenting with exotic ingredients, new flavors and textures, innovative preparation methods and new styles of presentation. Around the turn of the century, home cooks also partially returned to nostalgic dishes from their grandmother's era. At the same time, attention shifted to "healthy eating" in the form of organic and plant-based foods.

What Do We Eat Today?

The major changes that have shaped our diet since World War II are clear. "European agricultural and trade policy, technological innovations, globalization, economic growth, new health ideals and changing social structures have triggered and fueled the transformation," concludes professor Yves Segers. But what has it actually led to? What story do our shopping carts tell today? Are notable shifts still happening?

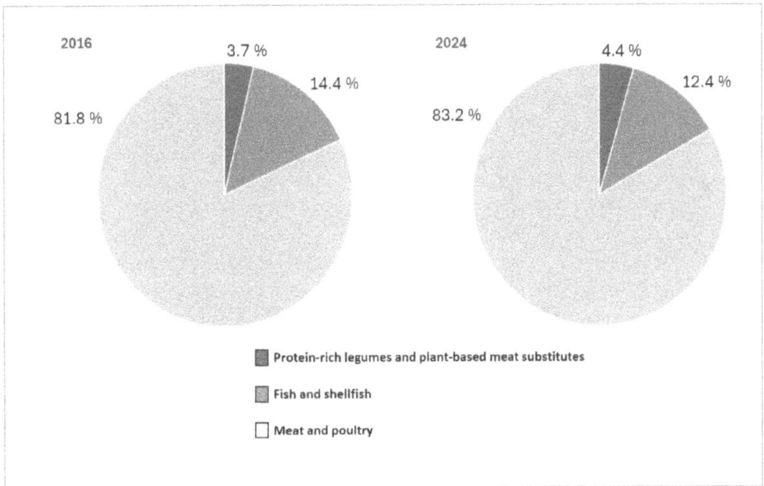

Figure 21: Purchases of meat and poultry versus fish, shellfish and plant-based alternatives in 2016 and 2024 (in percent). Source: YouGov Belgium via VLAM, 2025.

The meat–fish–veggie category tells the most interesting story. According to the latest report from the Flemish Center for Agricultural and Fisheries Marketing (VLAM, 2025), which consolidates several sources, we still buy more meat for home consumption (83% of volume) than fish and shellfish (12%) or plant-based alternatives (4%). These proportions have barely changed since 2016. At the same time, only half of Belgians say they eat meat four or more times a week. The share of households that say they alternate meat or fish with a plant-based alternative is rising cautiously: from 27% in 2016 to 36% in 2024. Supermarket purchasing behavior—especially of processed meat substitutes—seems to confirm this growing interest: in 2020, Belgians bought no less than 57% more meat substitutes than in 2014, though in absolute terms, volumes remain small compared with meat, poultry, fish and shellfish.

A Year in Our Shopping Basket

Proteins	
Dairy	77.7 kg
Dairy alternatives	8 kg
Meat and poultry	28.1 kg
Cold cuts (charcuterie)	10.4 kg
Ready-to-eat meat products and meat-based meals	5.6 kg
Fish and shellfish	4.2 kg
Plant-based meat and fish substitutes	0.6 kg
Protein-rich legumes	0.9 kg
Eggs	80 eggs
Fruit and vegetables	
Vegetables	44.6 kg
Fruit	40 kg
Starch	
Potatoes	24.5 kg
Pasta	5.7 kg
Rice	1.6 kg

Figure 22: An overview of average annual purchases per Belgian citizen across the main categories reported by VLAM. Source: YouGov Belgium, CPS GfK Belgium, VLAM Consumption Tracker (iVox) and VLAM calculations, 2023 and 2024.

Trend: Alternative Proteins

Rising interest in meat, fish and dairy substitutes is also reflected in the questions received by ILVO at the Food Pilot in Merelbeke-Melle. The Food Pilot is an experimental food processing plant equipped with the same equipment used in the food industry, but on a smaller scale. Every year this pilot plant welcomes more than 300 companies, from world-renowned players such as Barry Callebaut (chocolate) to relative unknowns like the start-up KarmaKarma (oatmeal cups). ILVO scientists advise these companies on new raw materials, innovative production processes, food safety, taste, shelf life and much more. Adjacent to the pilot plant is a lab complex where products leaving the line can undergo immediate testing.

The Food Pilot was opened in 2011 by the Flemish Minister of Economy to support innovation in the food industry. It is operated by ILVO and Flanders' FOOD, the spearhead cluster for the food industry. "In the early years, the sector faced the challenge of reducing salt, fat and sugar. Most tests and advice focused on that," explains communications officer Karen Verstraete. Later, the focus shifted to energy efficiency, use of by-products and preservation techniques. But a different trend is now emerging. "In 2024 we ran almost 500 trials in the production hall; half of them dealt with alternative proteins or dairy and meat substitutes. We've been seeing this for several years now, with requests coming from traditional meat and dairy companies as well as from raw material suppliers and innovative start-ups."

To meet this rising demand, ILVO and Flanders' FOOD invested in a new protein-processing line. It also includes equipment that can handle the small volumes typical of the rare, innovative raw materials being tested: microalgae, insects, the fermentation-based products from the beginning of this book and protein-rich plants. "It may sound simple to turn a protein-rich chickpea or alga into a tasty and healthy burger or spread, but it isn't. Companies still have many questions, and so do we. These range from the best way to pretreat or isolate the protein, to creating a pleasant mouthfeel with a crunchy bite, to ensuring food safety and shelf life. It's a new, fast-growing and extremely interesting field of research," says ILVO protein specialist Geert Van Royen.

Europe has long been self-sufficient in food and, as a producer, is on a par with much larger regions such as the United States or Asia. But in terms of protein, Europe is heavily dependent on imports—a growing challenge for the agrifood sector. Not so long ago, Belgian farmers grew faba beans and peas to feed their animals. Today, Europe imports millions of tons of cheap soy each year. In 2024 alone, that amounted to 15.7 million tons of soybeans and 18.7 million tons of soybean meal, mainly from South and North America. But relationships with these trading partners are under increasing strain, while questions about the climate impact of overseas soy keep mounting. For that reason, Europe has been investing in research on domestic protein crops for several years. Flanders joined in with its own protein strategy in 2021.

The Flemish Protein Strategy

In 2021, the Flemish government launched a plan to increase protein self-sufficiency by 2030. Among other things, the protein strategy calls for an expansion of acreage devoted to new plant-based protein crops for human and animal consumption, as well as research into new proteins such as insects, microalgae, microbial proteins and cultured meat. In its first phase, 76 actions and 42 research projects were initiated. Nearly half of these projects focus on innovative proteins, with another 22% focusing on plant-based proteins—unsurprisingly the same domains where the Food Pilot also receives the most inquiries.

Innovative Sources and Hybrid Products

At the Food Pilot, trials are underway with protein-rich plants as well as with microalgae and seaweed, fermentation products and by-products of food production. Legumes such as soy, chickpea, lentil, faba bean and yellow pea show promise, but so do (pseudo)cereals like quinoa and the press cake left over after extracting oil from sunflower or pumpkin seeds. Seaweed and microalgae are surprisingly versatile, offering flavor profiles that range from "fishy" and "crab-like" to "herbal" and "floral."

Microalgae are microscopic plants produced in a reactor. Hundreds of thousands of species are known, but only a handful are approved for consumption in Europe. Seaweeds are the larger types of algae; they can be cultivated either at sea or in open tanks on land. Both types of algae have a high protein content and an attractive amino acid profile, but also contain significant amounts of unsaturated fatty acids, antioxidants, minerals and vitamins. They have been eaten in Asia since ancient times, but they are slow to catch on in Europe because Western consumers and food companies are still unfamiliar with their taste. "If they are produced under safe conditions, you can actually do much more with them than you might think. We have already run successful experiments with algae in spice mixes, bouillon cubes, vegan mayonnaise, cheese, bread and dry sausage, but also in cookies, chocolate and liqueur," says ILVO scientist Johan Robbens.

Algae require little space and only need water, light and a suitable nutrient source to grow. That nutrient source can take many forms: among them, excess CO_2 from a dairy farm's pocket digester and nutrient-rich wastewater from a French fry producer. This efficient and potentially circular production process is something algae share with proteins from fermentation technology. Those, too, require few resources and little space, although more research is needed to make these processes truly circular and profitable.

A growing segment of the food market is hybrid products that combine animal proteins with plant-based or new proteins. These include burgers, sausages or meatballs containing meat in combination with vegetables, legumes, algae or oyster mushrooms. In 2025, Flemish pig farmer Mieke Verniest launched a "duo burger": 50% pork from the Flemish Ardennes and 50% split peas supplemented with carrot, ginger and turmeric. Attentive shoppers can now also find hybrid cheese and, more recently, hybrid milk. That milk innovation in Belgium comes from the Dutch dairy processor Farm Dairy and the Danish company Planet Dairy. According to them, hybrid yogurt and sweet desserts are next, hoping to reach the growing group of "flexitarians."

"For developing healthy and tasty vegetarian products, but also for hybrid forms with meat or milk, it's good to see more protein sources becoming available, as different sources can be better suited to the desired end result. All of the sources have their own amino acid profile, but they also have different flavors, colors and technical properties, which means they behave differently in end products. One protein foams better, another gels or binds better," explains ILVO protein specialist Geert Van Royen.

Challenges

Challenges for a successful protein transition take the form of cultivation and processing as well as nutrition and food safety. Scientists must seek a balance between food companies' enthusiasm about alternative proteins and the need for careful scientific study of possible health effects. "For dairy, meat, fish and eggs, we know most of the risks. We know what nutrients they deliver to our bodies and also what the downsides of consumption can be. For new and alternative proteins, we still need to unravel this. There is a knowledge gap," says ILVO scientist Els Van Pamel.

Proteins are made up of amino acids, which perform many essential functions in the human body. They are involved in building and maintaining cells and tissues, supporting the immune system and producing hormones and enzymes. They can also serve as an energy source. "There are nine essential amino acids that our bodies cannot produce themselves and that we must obtain from food. Depending on our age, we need more or less of them. A deficiency does not immediately show in weaker growth or resilience, but in the long term it can cause serious health problems," explains ILVO scientist and nutritionist Keshia Broucke. These include muscle loss, brittle bones, a weakened immune system, slower wound healing, anemia, depression, insomnia, fatigue and in severe cases, malnutrition disorders.

Meat, fish, milk and eggs all contain sufficient amounts of those nine essential amino acids. Among new and alternative proteins, currently only soy, quinoa and the microalga *Chlorella* have a complete amino acid profile, though this partly depends on cultivation and production conditions. Bioavailability and digestibility also vary: having a certain protein content in an ingredient does not necessarily mean our bodies will absorb it. "For instance, we know that some algae have an interesting protein profile, but the protein is locked inside a thick cell wall. If we eat them as they are, our body cannot access the protein. An extra processing step is therefore needed," Broucke notes.

This also applies to many plant-based proteins. They often contain anti-nutritional factors that hinder digestion and nutrient absorption. "By dehulling, soaking, cooking, heating, germinating, fermenting or grinding, we can often reduce these anti-nutritional factors, but a lot more research is still needed to optimize all these processing techniques and their effects on the human body. Post-harvest treatments and the variation in content across different varieties are also still under active investigation."

In other words, stay well informed, compose a balanced diet and above all diversify, so that your body gets all the nutrients it needs.

Amino Acid Score (≥ 1 is Sufficient) and Total Protein Content per 100 g for Various Protein Sources

	Amino acid score – animal sources					Amino acid score – alternative protein sources						
	Meat			Dairy		Legumes		Insects		Microalgae and seaweed		Microbial proteins
Essential amino acids	Pork	Beef	Chicken	Milk	Egg proteins	Lentils	Soy flour	Crickets (dried)	Mealworms (dried)	*Chlorella vulgaris*	Nori seaweed *Porphyra columbina*	Mycoproteins (dried)
Histidine	2.37	2.03	1.61	1.53	1.33	1.60	1.23	1.38	0.85	1.04	0.66	2.70
Isoleucine	2.06	2.03	2.32	2.41	2.38	1.82	1.66	1.63	0.76	1.59	0.97	3.00
Leucine	1.32	1.34	1.36	1.67	1.47	1.37	1.07	1.12	0.69	1.42	1.12	2.29
Lysine	1.72	1.81	1.77	1.48	1.15	1.39	1.08	0.87	0.45	1.15	1.04	1.80
Sulfur-containing AA	1.87	1.69	1.89	1.45	2.74	0.80	1.15	>1.23	0.60	1.01	1.43	1.82
Aromatic AA	1.36	1.41	1.33	1.75	1.74	1.52	1.29	1.63	0.71	1.37	0.99	1.68
Threonine	1.67	1.54	1.49	1.44	1.54	1.41	1.16	1.04	0.77	1.51	1.74	2.40
Tryptophan	1.28	1.20	1.28	1.37	1.64	0.97	1.07	0.92	-	2.09	0.57	2.40
Valine	1.84	1.71	1.69	2.10	2.52	1.70	1.38	1.48	1.13	1.89	1.67	2.70
Protein (g/100 g)	22.0	22.0	19.9	3.3	11.1	23.0	41.0	62.0	52.0	42 - 58	21.0	46 - 48

Figure 23: This table shows the Amino Acid Score of animal, plant-based and novel proteins for each of the nine essential amino acids. A score of 1 or higher indicates that the protein source provides a sufficient amount of that essential amino acid. The last row shows the total protein content per 100 g of product. Some sources with a high total protein content are deficient in some essential amino acids. Source: ILVO, based on various scientific publications; see bibliography.

Food Safety

A key concern with novel protein sources such as algae, insects and microbial proteins is their chemical and microbiological safety. Duckweed and seaweed, for example, are known to sometimes accumulate heavy metals. Using certain fungi to produce microbial proteins can also carry risks due to the possible presence of toxins–so-called mycotoxins–which can cause both acute and chronic health problems in humans and animals. Nor do we know the extent to which these new or alternative raw materials are contaminated with microorganisms and how to eliminate them via heat treatment. Take oats, the raw material for oat milk. Oats have an entirely different microbial load and composition than raw cow's milk. At what temperature and for how long should oats be heated to produce a safe, shelf-stable and tasty product? "We hope to answer these questions in the coming years," says Els Van Pamel.

The good news is that research into novel proteins is advancing at lightning speed. There is every reason to expect that knowledge gaps will be closed quickly and that companies working with new proteins will soon be able to adjust their production systems accordingly. "It is already encouraging that tomorrow's diet will offer more variety and choice of proteins. Alongside meat, fish and dairy, we are getting a whole range of plant-based and novel proteins. That gives us greater freedom to diversify. In the end, diversification will be part of the solution to many diet-related health problems."

Life Cycle Analysis

But what about the ecological sustainability of novel and plant-based proteins? To calculate this properly, the Food Pilot installed energy sensors on every piece of equipment, old and new. "The carbon footprint of animal-based products is already well documented. The greatest impact occurs in the earliest stages of production: on the farm and in feed production, especially when using imported soy. For plant-based products, however, data is still lacking. Their impact at the production stage is smaller, but later processing is often more complex. Seeds must be dehulled and heated, algae and microbial proteins need to be dried, the proteins often need to be extracted and then texturized and so on. We are studying the contribution of each step to the total climate impact so that policymakers and consumers can make informed choices," explains ILVO expert Veerle Van linden.

When by-products are used as alternative protein sources, it is equally important to assess the full life cycle of the products in order to avoid shifting problems elsewhere. "If you take the starchy by-products from French fry factories and process them into new food products, they can no longer be used as animal feed. How will those starches be replaced in livestock systems? What knock-on effects can we expect? As scientists, we are trained in systems thinking to analyze these complex processes and transitions thoroughly. We apply it across all agricultural, climate and food-related challenges."

Systems Thinking

Systems thinking is a way of approaching complex environments or challenges—like in the food system—in an interdisciplinary manner, with attention to the many interactions between interlinked components and aspects. Only in this way can we create sustainable change without the risk of trade-offs or problem shifting. ILVO actively trains its researchers to look at the bigger picture in an integrated way.

Investments

To accelerate research into new proteins and protein sources, ILVO has invested not only in a protein-processing line but also in a pilot hall for "post-harvest processing" of crops. The Post Harvest Pilot in Merelbeke-Melle can process seeds, grains, flowers and entire plants. Drying, sorting and cleaning are all available according to the requirements of a specific application. These may be intended for food—organic or otherwise—but equally for raw materials for building materials, paint, essential oils or even medicines. "What is unique about this new pilot hall is that it can handle small volumes. There was real demand for that in Flanders, where our agribusinesses have large-scale drying and processing facilities, but farmers with a new, innovative crop had nowhere to go with their small harvest," explains Kristiaan Van Laecke, head of ILVO's Plant Sciences department.

In Ostend, ILVO has also built a specialized aquaculture hall to study fish, shrimp and algae grown on land. Back in Merelbeke-Melle, a new Feed Pilot supports research into subjects such as the inclusion of by-products and new protein sources in animal feed. "With these investments, we aim to close the protein innovation loop and remove barriers so that the entire agrifood chain can adapt to the new reality in which we will produce more domestic and alternative proteins. We are ready to guide agriculture, aquaculture, the food industry and the feed sector through this transition," says Bart Sonck, head of ILVO's Animal Sciences department.

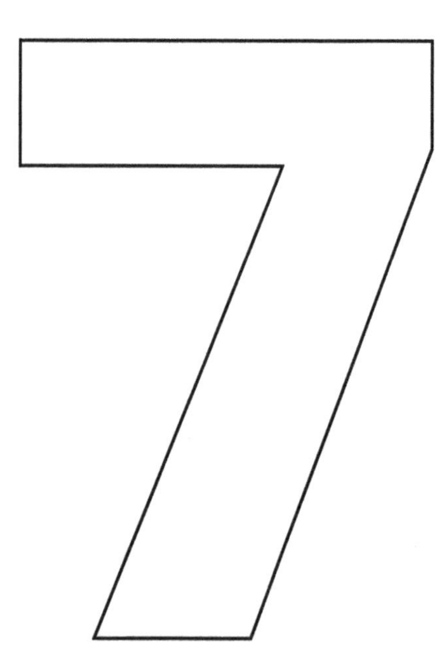

CHAPTER 7

It's Time to Roll Up Our Sleeves and Get to Work

Reflections by Joris Relaes, Administrator-General of ILVO

The year 2022 was a festive one at ILVO. We celebrated our 90th anniversary with a multi-day festival "of food and facts." Ten thousand curious visitors—scientists from home and abroad, local residents, partners, clients and school groups—flocked to Merelbeke-Melle to join the festivities. Between lectures and lab tours, a panel discussion was held with past and present directors. They marveled at how quickly the world, agriculture and ILVO itself had changed. What raised the most eyebrows? In its early years, ILVO received funds from the American Marshall Plan to speed up Europe's reconstruction after World War II. With those American dollars, ILVO researchers developed machines to clear forests and drain marshes. Barely a generation later, ILVO is researching agroforestry and rewetting, and 6 of its 240 hectares of trial fields are being converted to nature. Several hedgerows, a pond and a few hundred trees have already been installed.

Times change. The agrifood sector has also transformed rapidly, and the pace shows no signs of slowing. Canned goods are losing popularity while ready-to-eat meals—in every possible form, often delivered straight to the doorstep—are booming. Despite the proliferation of cooking shows and cookbooks, we eat out more and cook less at home. But the food industry has found ways to respond to that too.

In the previous chapters, together with ILVO scientists, we have looked at the future and the role of research in the rather optimistic title of this book: *Our Food Will Be Fine*. At the same time, I believe we should pause and reflect on the broader context in which these technologies and trends are unfolding. Globally, four major forces have shaped the agrifood sector in recent decades: innovation has driven primary production; globalization has advanced rapidly; large private players have assumed key roles; and our eating habits have shifted. These forces will evolve further, encouraging the agriculture, fisheries and food sectors to lean in a particular direction.

Trend 1: Innovation Drives Primary Production

The question of food is as old as humanity itself. Even though our Northwest European climate is relatively mild and our soils more fertile than many other regions, even the West has not always been able to produce enough food. The many wars, floods and epidemics often caused famines. Excavations of medieval skeletons show that nearly all population groups experienced varying periods of undernourishment. There were also years of plenty, but everything was framed in the biblical belief that "the Lord giveth and the Lord taketh away." Only the scientific approach of the late 19th and early 20th centuries contributed to a substantial growth in agricultural productivity. First came Mendel's laws of inheritance in plants and animals—the foundation of breeding and the development of improved varieties. Then came the Haber-Bosch process, which gave us artificial fertilizer. After World War II, crop protection products and antibiotics were added to help control pests and diseases in plants and animals.

Today, as you have read here, we again stand on the brink of major technological breakthroughs: advances in CRISPR-Cas, precision fermentation, artificial intelligence, robotics, big data and smart sensors. These hold the promise of more precise approaches, more efficient use of resources, and a smaller climate and environmental footprint.

Innovation has already relegated mass famines in the West to the history books. We do not often realize how extraordinary it is that this planet manages to produce enough food to sustain eight billion people. I am convinced that innovation will also provide answers to the environmental and climate challenges of food production. It will deliver stronger plants that will receive precisely the amount of water, nutrients and protection they need—no more, no less—which will drastically reduce the impact of food production on biodiversity and climate. It will give us smaller, electrically powered robots instead of ever-larger and ever-heavier machinery, thus sparing our fertile soils. Is that the reality on the field today? Not yet. But robot prototypes are already driving on ILVO's experimental fields, and AI models that predict pests and diseases are improving at lightning speed as they are trained on both real and synthetic data. Just as solar panels and robotic lawnmowers have become accessible to ordinary households, robots and AI models will in time become standard equipment on the average farm.

Trend 2: Globalization

The food trade is nothing new. The Romans traveled to Egypt to source grain from the fertile Nile Delta. In the Middle Ages and during the era of merchant fleets, trade in (mostly non-perishable) agricultural and food products flourished. By the late 19th century, American grain was crossing the Atlantic by steamship, flooding the European market. But the past five decades have seen a surge in globalized agrifood trade. In that period, world trade in agricultural and food products has increased fivefold, thanks to improved transport and storage methods, the establishment of the World Trade Organization (WTO) and the global reduction of import tariffs. Sharp rises in foreign investments have further accelerated globalization. Try to find a city today without a McDonald's, Burger King, or Starbucks in the neighborhood.

Yet, of all agricultural products produced worldwide, only 10% are actually traded on the global market. Contrary to popular perception, most of our food is still consumed locally. This is due not only to the short shelf life of many products, but also to their relatively low value compared to their weight or volume. Economically, it makes far more sense to trade sugar or frozen fries than the raw sugar beets or potatoes.

Slowbalization

Since the Covid-19 crisis and Russia's invasion of Ukraine, we live in a world marked by rising geopolitical tensions. The era of rapid global trade expansion that began in 1995 with the launch of the World Trade Organization has now come to an end. According to *The Economist*, we are now in a phase of "slowbalization:" not outright contraction, but stagnation of global trade. There is a growing push for strategic autonomy.

We first experienced how these geopolitical tensions could impact agriculture in 2014. In response to European sanctions over the annexation of Crimea, Russia imposed an import ban on a wide range of European agricultural and horticultural products. That was an uppercut for Belgium: at the time, Russia imported 21% of all Belgian agricultural products. Almost overnight, pears, tomatoes and pork suddenly had to find new markets. Trade with China is also proving difficult. And the Flemish potato processing sector recently faced major export challenges to Brazil due to trade disputes. Closer to home, Brexit has complicated trade with the United Kingdom. The age of ever-expanding free trade is over.

So how can the export-oriented Flemish and European agrifood sector adapt? Risk diversification—avoiding putting all of our eggs in one export basket—seems to me to be the most obvious strategy. We are also seeing some companies invest abroad and partially relocate their production, in part to secure the supply of raw materials. Spurred by lessons from Covid and ongoing geopolitical instability, companies are mapping out vulnerabilities in their supply chains. What alternatives exist for their most critical ingredients without compromising on taste, functionality, or nutritional value? Some are even exploring entirely new ingredients. Several chocolate makers, for example, are experimenting with cell-based or microbially produced cocoa as an alternative to the increasingly expensive cocoa beans.

At the same time, it seems wise to refocus more on the internal European market, which still counts a large number of affluent consumers who value sustainability and quality, areas in which Flemish products excel. There may also be opportunities in exporting to Africa, where populations and food demand are growing. South Africa, for example, has recently entered the top 10 of distant markets for Belgian food exports. That said, reciprocity in trade relations will be essential, ensuring that African products also gain easier access to European markets. Trade need not be a zero-sum game in which one side loses when the other gains. When markets are organized fairly, everyone can win.

Trade Conflicts and Agreements

Donald Trump's second term as U.S. president promises to be marked by trade conflict with Europe. The main sticking point? Agricultural products. But here too, there's nothing new under the sun. From the very beginning of European unification, agriculture has been a disruptive factor in transatlantic trade relations. It began with the infamous "chicken war," sparked when cheap American frozen chicken flooded the European market in the 1950s and 1960s. Europe responded with higher import tariffs, to which the U.S. retaliated with increased duties on cognac, starch and light trucks. The first was an important French export, the second a Dutch one, and the third? You guessed it—a classic German export. The economic impact was quite limited, but the stage was set for decades of alternating tension and détente in trade relations. While most tariffs have disappeared over time, the "chicken tax" on light trucks has never been rescinded. It is the main reason why nearly all pickups on American roads are made in America. The episode also gave rise to the famous remark by German Chancellor Konrad Adenauer about his correspondence with U.S. President John F. Kennedy: "In the past two years I have exchanged many letters with President John F. Kennedy about Berlin, Laos, the invasion in Cuba... but I guess that about half of them have been about chickens."

European farmers too have long been wary of opening borders to other countries or trade blocs. Yet history has often proven that we benefited more from increased trade than we lost. The Flemish horticulture sector trembled with fear when Spain and Portugal joined the European Union. Many wondered whether our tomato and strawberry sectors would survive. In reality, the opposite happened. We adapted by developing new varieties, and today Flemish tomatoes and strawberries dominate supermarket shelves more than ever. The dire predictions of an economic bloodbath from enlarging the EEC to Central and Eastern Europe never materialized. And the free trade agreement with Canada (CETA), which Belgium's southern region Wallonia tried to block until the very end under Paul Magnette, did not spell disaster for either European or Belgian farming. On the contrary: recent European Commission reports show that European agriculture has actually gained more from the agreement than the Canadians have.

There is no doubt that we must push for a level playing field in any trade deal. But we must always keep the bigger picture in mind. There is still resistance within Europe to a trade deal with the Latin American Mercosur countries (Brazil, Argentina, Uruguay and Paraguay), but in today's turbulent geopolitical climate, such an agreement cannot simply be dismissed. It would almost certainly boost the European economy and, with it, the purchasing power of European consumers, who would in turn buy high-quality Flemish farm products. Better the clear framework of a trade agreement than the arbitrariness and opportunism of current U.S. trade policy.

Where Is Our Food Produced?

This raises two key questions: Where is what produced? And which countries are truly self-sufficient? The answer may surprise you. A recent article in the journal *Nature Food* shows that only one country on Earth is capable of fully providing its population with a healthy, varied diet: Guyana, a South American nation of about 800,000 people which also happens to be the fastest-growing economy in the world since their discovery of massive offshore oil reserves. Every other country in the world depends on trade to feed its population healthy food, even the countries that produce more calories than they strictly need.

Broadly speaking, there are five categories of countries in the global food market. First are the so-called "established food producers" such as Belgium. These are countries with fertile land and favorable climatic conditions for farming that underwent an early agricultural revolution (mechanization) and, in general, produce more food than they consume. This group includes the United States, Canada, Australia, New Zealand and much of Europe (not only Belgium but also Germany, France, Poland, Spain and the Netherlands).

European Strategic Autonomy

I do wish to make an important caveat for Europe. In two critical areas, we are not self-sufficient. First, plant-based proteins. The textbook example is soy, which we import in massive volumes from both North and South America. Second, mineral fertilizer. We remain highly dependent on imported fertilizer, including from Russia. In these uncertain geopolitical times, it is essential that we strengthen Europe's degree of self-sufficiency on both fronts. One way forward is to adjust our crop plans to give more space to alternative plant-based protein sources. Efforts to introduce soy, peas and faba beans into mainstream farming are a first step in the right direction. But more must be done to trigger a fundamental shift in the crop planning of Flemish arable farmers. Why not consider a short-term but strong financial incentive from the European agricultural budget to stimulate cultivation of these crops?

As for imports of mineral fertilizers, the solution lies in circular thinking. We need to recover more of the plant nutrients already produced in the livestock sector or found in the by-products of the food industry, and reuse them as tailored fertilizers for arable and horticultural crops. If we succeed, we will be one step closer to the strategic autonomy in agriculture and food production that Europe urgently needs.

United Kingdom

One Western country is not found on the list of the group of established food producers: the United Kingdom. It imports about 40% of its basic food from abroad. This was already the case before Brexit; it is a historical artefact of the British Commonwealth. The British were simply accustomed to easy provisioning from their former colonies and, therefore, invested little in their own food production. The lords preferred hunting in their vast estates and forests to managing their own farms.

A second category of countries that play an important role in global food trade are the fast-growing "new food producers." Consider Brazil, which in the 1970s was still a net food importer but has since become the world's leading exporter of beef, chicken and sugar. Other members of this group include Uruguay, Paraguay, Thailand and Vietnam, as well as Russia and Ukraine. They also have favorable conditions for farming, but their agricultural revolutions began later than those of the established food producers.

Russia and Ukraine

Both Russia and Ukraine saw their agricultural exports increase sharply in the years before the war. The Russian invasion of Ukraine generated real concern for disruption of the global food supply. It was difficult to explain why this would not cause the price of bread at home to double. After all, in 2022, Russia and Ukraine together accounted for nearly 30% of global wheat trade. But that did not mean they produced 30% of the world's wheat or covered 30% of global demand. The partial loss of Ukrainian grain exports did not cause major shortages on the world market, and certainly not in Europe, which produces as much wheat as China. The only real effect was a bad case of nerves, which drove prices up.

A third group of countries produces just enough food to feed their growing populations. Typically, a high percentage of the workforce is employed in agriculture, primarily on small farms. These are often countries where governments have invested heavily in rural development and have shielded their domestic markets from global competition. Examples include Côte d'Ivoire, Guatemala, Turkey, Pakistan, Bangladesh and Indonesia, but also China and India. The latter two, each with populations of over a billion people, have undergone a spectacular agricultural transformation since the Green Revolution and the breeding of wheat, rice and maize since the 1960s. Mexico and Indonesia have also seen a spectacular rise in grain yields.

Norman Borlaug and the Green Revolution
The Green Revolution, often referenced throughout this book, was driven by the groundbreaking work of American plant breeder Norman Borlaug. His improved grain varieties saved hundreds of millions of people from malnutrition and famine, and as noted above, he received the Nobel Peace Prize for his efforts. Criticism of Borlaug's model only came later, particularly after the 1972 publication of *The Limits to Growth* by the Club of Rome, which highlighted the reliance of his approach on intensive use of pesticides, fertilizers and water.

Food-Insecure Countries and Importers

The fourth and fifth groups of countries are not self-sufficient. They often have little fertile land, limited water resources and a high population density. Oil-rich nations such as Saudi Arabia, Qatar, the United Arab Emirates and Kuwait belong to the category of "wealthy food importers." Japan, South Korea and Singapore also fall into this group. They have enough purchasing power to compensate for their lack of domestic production but are now seeking to boost their own food autonomy. The most striking example is Saudi Arabia, which has created irrigated agricultural zones in the middle of the desert.

A large group of other countries lacks both the purchasing power and the natural resources to feed their populations. This includes many nations in Central America, West and Central Asia, North Africa, the Middle East and Sub-Saharan Africa. Strikingly, the latter region was still self-sufficient in 1970, but its population has grown far faster than its agricultural production. Egypt is a case in point: in 1960 it was self-sufficient, but today, with four times the population it had then, it is the world's largest wheat importer.

Mexico

Remarkably, Mexico has belonged to the list of "food-insecure countries" since 1980. It is astonishing that the birthplace of maize—and one of the countries transformed by the Green Revolution—is no longer self-sufficient today. In fact, it has become the world's largest maize importer. The free-trade agreement with the United States, which opened the floodgates to large volumes of cheap American maize, is one reason. More importantly, however, Mexico neglected its own agriculture and rural areas for too long. As a result, more than 75% of Mexicans now live in cities. Mexico City, with over 20 million inhabitants, is among the largest cities in the world and, consequently, has to deal with typical megacity challenges, not the least of which is drug-related violence.

Trend 3: Heavy Market Concentration

Back to the countryside. Worldwide, there are an estimated 600 million farmers, including 500 million subsistence farmers who produce only enough food for their own families. This immense number contrasts sharply with the concentration of power among input suppliers and food processors. This is the third major global trend. More than half of all agricultural and horticultural seed production is controlled by just 10 companies. In crop protection, the concentration is even greater: 10 companies dominate 80% of the global market.

A similar pattern appears in the first stages of processing and trading of products such as grains and oilseeds, where the so-called ABCD companies—Archer-Daniels Midland, Bunge, Cargill and Dreyfus—hold sway. Concentration in distribution and retail is also evident, although somewhat less pronounced. Think of Walmart, Tesco, Carrefour, Lidl and Aldi. In contrast, companies producing finished food products are generally less consolidated. This is certainly the case in Flanders, where the food sector still consists largely of small and medium-sized enterprises.

This stark difference in the number of companies across the different links of the agrifood chain results in an equally stark imbalance in power and negotiating strength. Think of an hourglass with millions of farmers at the top and millions of consumers at the bottom. They meet via the narrow middle, which is controlled by a handful of companies with market power. These firms largely decide what ends up on our plates and at what price. The hourglass metaphor is not perfect, but it illustrates the outsized influence of suppliers and buyers in the agrifood chain.

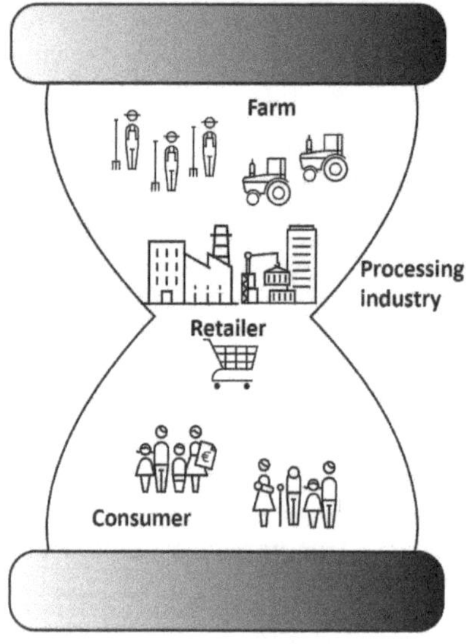

Figure 24: The hourglass model as applied to the agrifood market. Free translation and illustration by ILVO.

Trend 4: A Changing Diet

A final striking global trend of the past 50 years is that we are all eating both more and differently, just as we saw in the previous chapter. Alongside the general rise in prosperity, our diets have undergone "meatification." From diets once based mainly on plants, we have shifted toward higher consumption of animal proteins, sugars, oils and fats. And this meatification has not yet peaked. The European Commission expects global meat consumption to rise by another 41 million tons by 2033. It's hard to imagine, but that is the equivalent of Europe's entire current annual meat production. The type of meat we are eating is shifting, however: poultry has become the new favorite. The ecological and climate footprint of chicken is significantly lower than that of pork or beef, mainly because chickens require less feed to produce the same amount of meat. It takes 1.6 kilograms of feed to produce 1 kilogram of chicken meat, compared to about 2.8 kilograms for pork and 8.7 kilograms for beef, depending on the breed. At the same time, meat production has risen proportionally with consumption. Between 1950 and 2005, the world population "only" doubled, but meat production increased fivefold. In China, the shift was even faster, with meat production increasing fivefold in just 25 years.

Meat and Prosperity

The world is full of vegetarians—not out of conviction, but out of necessity. Meat consumption is closely tied to income, with a general rule of thumb: if people earn less than $1,000 per year, they eat mostly plant-based diets. Once incomes rise to between $1,000 and $5,000, meat consumption increases exponentially. Above $5,000 per year, it levels off. China is a prime example of a country that has gone through this transition. In 1990, its GDP per capita was $350; today it exceeds $10,000. While its population has doubled over the past 50 years, meat consumption has increased fivefold. Yet even so, China's meat consumption—about 60 kilograms per person per year—still lags behind that of the United States, where the average is 110 kilograms.

The global consumption of convenience foods has also risen dramatically. Since 2008, more people worldwide live in cities than in rural areas, and urban dwellers increasingly opt for ready-made meals or eating out. In some cities, like London, apartments are already being built without kitchens, based on the assumption that a microwave is enough to reheat delivered meals. In response, other urban movements have sprung up that advocate for slow food, raw food, organics, or short food supply chains—a return to authentic, pure products and direct contact with producers. For now, these remain niche markets.

Pathways Forward

Global food consumption is expected to keep rising for some time. It will be a challenge to feed a projected world population of 10 billion by 2050. Yet the bigger challenge for agriculture—including in Flanders and Europe—is not whether we can produce enough food, but how we can do so sustainably. How do we provide 10 billion people with a healthy, diverse diet within the ecological limits of our planet?

One obvious solution that often receives too little attention is reducing food waste. What could be more economically and ecologically efficient than actually eating all the food we produce? In practice, this remains a sticking point. As we mentioned above, it is remarkable that we succeed in supplying our cities with plentiful, healthy and diverse food while so much still goes to waste. The UN Food and Agriculture Organization (FAO) estimates that one-third of all food produced is never consumed. In many southern countries, more than 40% is lost after harvest or during processing due to poor storage and transport conditions. In Western countries, the major losses—also estimated at 40%—occur further down the chain, at the levels of distribution and in consumers' homes. It's up to all of us to tackle this together.

In addition, we need a new generation of Norman Borlaugs and a new Green Revolution, one that combines the foundational knowledge of the past with today's technological possibilities. The scientists featured in this book have shown us potential pathways forward. They see the beginnings, but no one knows yet where those paths will lead and how they will evolve. That uncertainty is uncomfortable and typical during times of major transformation. But we must choose the way forward.

There is also a growing call for an all-encompassing agricultural and food vision, fueled in part by the farmers' protests in Belgium and the Netherlands in 2024. Among their many complaints was the European agricultural policy which, despite several reforms, has failed to deliver on its ecological promises. It remains difficult to achieve a number of Europe's environmental and nature objectives because, in a globalized economy, it is equally difficult to maintain a level playing field. How do you ensure that European farmers invest in resource-efficient technologies when their counterparts elsewhere do not, yet are still able to sell their products in European supermarkets, often at lower prices? This is not only about ecological sustainability but also about food safety, quality and animal welfare. In response to the large-scale farmers' protests of 2024, policymakers made a political commitment to develop a comprehensive agricultural vision, both at the European and the Flemish level. Perhaps they could take inspiration from Denmark and Sweden? These two countries have already done their homework and—independently of one another—launched two very different but equally valid visions.

The Danish Model

In Denmark, a large majority of the government and the opposition approved a new agricultural and food policy in parliament. The Danes believe strongly that their export-oriented sector deserves every chance it can get. They see drastic sustainability measures not as a threat but rather as a chance to strengthen their competitive edge. They intend to leverage "intensive sustainability" as a selling point in export markets, and to compete head-on with other global players. The Danish plan also provides for converting 15% of farmland into nature, planting one billion trees and introducing a carbon tax. The government has earmarked €5.7 billion for this, and the private sector is also contributing. Pharmaceutical giant Novo Nordisk, known for the weight-loss drug semaglutide (marketed as Ozempic and Wegovy), is putting in another €1.6 billion.

The Swedish Model

Sweden has chosen a very different agricultural policy, as it also starts from a very different baseline. The country is only about 50% self-sufficient and wants to reduce its dependence on imports. That urgency has been heightened by geopolitical tensions with its hostile neighbor Russia. The Swedish plan includes a range of fiscal incentives for the farming sector, measures to make jobs in the food industry more attractive and the protection of farmland.

The Flemish Model?

Like the visions developed in these two Scandinavian countries, a Flemish agricultural vision must also start from its own specific context. What are our strengths? Flanders has ideal production conditions: a temperate climate with mostly rain-fed farming, fertile soils, highly trained farm managers and a strong supply and processing industry. It is no coincidence that the food and beverage industry has consistently proven itself to be a key driver of the Belgian economy. It is the country's largest industrial sector, accounting for roughly one-quarter of total industrial turnover, 14% of Belgian exports and by far the largest number of industrial jobs.

Another major asset is our ports, which make it easy to import raw materials and export finished products. Zooming out on the map, one could even say that we farm in a harbor. The sector is also surrounded by universities, research institutes and practical research centers that provide ample opportunities for continued innovation. And last but not least, the sector is located in the middle of an area filled with highly affluent and quality-conscious consumers. This does not mean there are no challenges. Ask the farmers themselves, and they will tell you that access to affordable, high-quality farmland is one of the biggest. Flanders, like the Netherlands, is densely populated and heavily urbanized, and much of it is paved. By definition, farmland is limited and lies under heavy pressure from competing demands for housing, industry, nature and recreation. Expanding the amount of farmland is not a viable path to sustainability, as it was 80 years ago. Instead, we must manage the existing land more sustainably and maximize investments in what could become a new Green Revolution. Every link in the Flemish agrifood chain must be prepared to invest in that future, and therein lies the problem.

The Family Farm?

This brings me to one of the current challenges for Flemish agriculture: the resilience of the traditional family farm. Thanks to the agility and flexibility of this earning model, Flemish farming has survived many real and perceived crises. Recovering from an outbreak of swine fever or from an energy crisis in greenhouse horticulture is not a 9-to-5 job with overtime pay and an end-of-year bonus. But today the family-based approach is under severe pressure. The demands imposed by the market and the government are almost impossible to bear for the traditional family farm. Besides their actual farming activities, farmers must manage taxes, staff, marketing and quality and sustainability monitoring, all of which require time and expertise. The question is whether all of this can still rest on the shoulders of a family farm. Expertise can be hired, of course, but that comes at a price.

Traditional Farm?

Contrary to popular belief, the family farm is not a centuries-old hallmark of Flemish agriculture. The traditional family farm with about one-and-a-half full-time equivalents only really emerged during the Interwar period, especially when tractors replaced horses, a transition that accelerated dramatically after World War II. More than a century ago, Flanders had few family farms. Instead, there were large estates run by a small army of hired help: stewards, farmhands, stable boys, horse handlers, seasonal workers, day laborers and domestic help.

Two options remain: either farms grow into larger enterprises with specialized departments focusing on different aspects of running the farm—we already see this happening in greenhouse horticulture—or we look for creative forms of cooperation that can relieve family farms of part of their burden. I am aware this is a sensitive issue that requires broader social debate. It entails a loss of independence and triggers painful memories of unfair integration practices by supply chain partners in the 1980s and 1990s. But why not think in terms of a franchise model, in which farms retain autonomy over production decisions while relying on a central service provider for other aspects? The cooperative model could also be reimagined. Why couldn't retiring co-op members make their land available to younger co-op members who wish to build a future but who cannot afford to buy land?

Of course, there is also a third option. Alongside larger enterprises, cooperatives, or franchise-style farms, there will continue to be farms that sell directly to local consumers. These are usually smaller operations that devote a significant share of their time to running a farm shop, a pick-your-own operation or a vending machine, offering educational activities or agritourism, or even hosting children's parties—you name it. That too is a path forward, though not for every farm in Flanders. Ideally, this question should be considered in a spatially tailored way: which farming model makes sense where—what crops, what livestock, what side activities, for how many potential customers?

The Final Word: Our Food Will Be Fine

Let me say it again: it is nothing short of a miracle that, as a society, we succeed in supplying cities like London with three varied meals a day, every day, as Carolyn Steel writes in her brilliant book *Hungry City* (2008). This requires not only sufficient food production but also a highly efficient distribution system. History has repeatedly shown that such a system can only work in a free market, but also that the free market has its limits. We cannot close our eyes to the challenges: climate change, biodiversity loss, ethical questions about animal welfare, farmers feeling like hostages of a hostile socio-economic system, skewed power dynamics in the food chain, water availability and the growing obesity crisis across large parts of the world.

A list like this is enough to make anyone feel pessimistic. But there is only one way forward. To put it in farming terms: we must "put our hand to the plow" and move ahead. We cannot waste too much time debating our visions for a future decades away. We need to roll up our sleeves and get to work today.

That is also ILVO's motto. We continue to work toward greater sustainability in agriculture, fisheries and food—economically, ecologically and socially. But we are aware more than ever that constant adjustment is needed. Louise Fresco, Chair of Wageningen University & Research until 2022, said it perfectly in a 2025 opinion piece in the newspaper *NRC*: "Progress is not a straight line of perfect solutions, but a skating movement that continually requires correction." That reminds me of our electric cars. If we had waited until we were completely sure that their ecological footprint was watertight, we would still be debating today and still driving almost exclusively on fossil fuels. Thankfully, automakers did not wait for the conclusion of that debate. They released electric cars that, after some trial and error, now have a much smaller ecological footprint than just a few years ago.

This approach is also the guiding principle throughout ILVO's research, as you will have noticed while reading this book. There are still questions about the impact of alternative proteins and microbially produced fats, but we are already making progress and answering questions along the way. Robots and AI models are still costly and imperfect, but we are already building prototypes and sandbox environments to push progress forward. The 750 ILVO employees working in our 90-year-old organization are still motivated to search for solutions. We combine core values from the past with the newest technologies. We go back to basics, not with one eye on the rearview mirror, but in tandem with today's innovations.

That is why I wish to stay hopeful. In this book, ILVO scientists have touched on some of the technologies they are developing or the trends they see emerging in their data. Real people are working on real solutions—and not only at ILVO, either. Research is increasingly a global enterprise. Through collaboration in many mainly European projects, our researchers are in near-constant contact with colleagues across Europe and beyond. The digitalization of society has made data instantly available everywhere. That too can turbo-charge innovation.

I do not want to sound naïve, but I am sure our food will be fine. I like to quote the Danish agricultural economist Ester Boserup: "Necessity is the mother of all invention." We will rise to our challenges if we bundle our strengths and all take our own responsibility. By "we" I mean researchers, of course, but also suppliers, farmers, processors, supermarkets and you as a consumer.

Acknowledgments and Peer Readers

This book was made possible thanks to the input, careful review and support of colleagues and scientists at Flanders Research Institute for Agriculture, Fisheries and Food (ILVO). ILVO researchers and research support staff shared not only their time but also the hidden knowledge within our organization. That knowledge is often passed along silently and under the surface. In this book, we drew it to the surface and put it under a spotlight. The tacit knowledge bundled in this book cannot be found in our project reports, on our website or in our PowerPoint slides.

Explicit and Tacit Knowledge

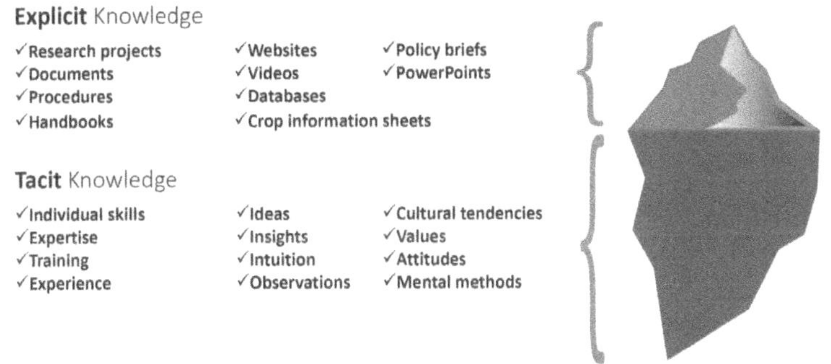

Figure 25: Metaphor of the iceberg applied to two different types of knowledge. Freely translated and illustrated by ILVO.

The process of writing and reviewing was itself an exercise in systems thinking. The greatest strength of this book lies not in a single statistic or some fantastic discovery, but rather the overall picture it paints of three sectors in transition: agriculture, fisheries and food. If the book succeeds in offering a taste of that to curious consumers, policymakers or influencers, then we will have reached our goal.

Above all, we hope it resonated with you and will linger in your mind for a while.

Many thanks to Joris Relaes for his broad perspective and trust in the writing process. Many thanks to Nele Jacobs for the synthesis and writing. Many thanks to Sandra Leroy, Greet Riebbels and Nancy De Vooght for their support and critical reflections. Many thanks to Miriam Levenson for this English translation. Many thanks also to the Center for Agrarian History and its coordinator, Professor Yves Segers, for providing input and contributing a section on our changing food culture. Many thanks to the companies and organizations who allowed us to cite them here. And many thanks to all ILVO scientists listed below who shared their knowledge and engaged in dialogue with each other.

Peer Readers

- *Bernaert, Nathalie: Food technologist with expertise in plant-based processing and the valorization of food by-products*
- *Blondeel, Lancelot: Researcher in sustainability and digitalization of Flemish fisheries*
- *Broucke, Keshia: Food researcher with expertise in hybrid meat products and plant-based protein sources*
- *Cougnon, Mathias: Plant breeder and crop researcher*
- *Cool, Simon: Research engineer with expertise in precision farming, computer vision, robotics and sensor technology*
- *D'Hose, Tommy: Soil expert focused on carbon storage, soil quality and soil compaction*
- *De Boever, Maarten: Soil expert with experience in organic farming and water management*
- *De Campeneere, Sam: Researcher in sustainable cattle farming with expertise in emissions–Scientific Director*
- *De Grande, Annatachja: Engineer in food and feed technology, nutritionist and researcher in feed applications–Feed Pilot Coordinator*
- *De Keyser, Ellen: Coordinator and communications officer for ILVO's Plant & Soil Living Lab–researcher with a biotechnology background*
- *De Loose, Marc: Plant biotechnologist with an interest in the relationship between food and health–Scientific Director*
- *De Swaef, Tom: Plant physiology researcher and plant modeling expert*
- *De Temmerman, Pieter-Jan: Researcher in precision livestock farming and applied AI*
- *De Waegemaeker, Jeroen: Researcher in spatial policy and climate adaptation*
- *Delacauw, Sander: Researcher in applied AI and image recognition for marine monitoring*

- *Derycke, Sofie: Marine biologist and molecular ecologist*
- *Facq, Ennio: Researcher in carbon farming and systems analysis*
- *Feyaerts, Dylan: Coordinator of ILVO's Living Lab for Agroecology and Organic Farming*
- *Fosselle, Sylvie: Social researcher with a focus on agriculture and society*
- *Garré, Sarah: Researcher on water and agriculture with an interest in science communication–Director*
- *Goossens, Karen: Researcher in sustainable cattle farming*
- *Herman, Lieve: Food microbiologist with expertise in food safety–Head of Department*
- *Heungens, Kurt: Expert in plant health and plant-pathogenic fungi*
- *Hostens, Kris: Marine ecologist researching the exploitation of and impact on the marine ecosystem*
- *Hunninck, Marijke: Project manager and bridge-builder in innovation projects on agriculture, data and technology*
- *Lambrecht, Ellen: Molecular microbiologist with expertise in food safety*
- *Malengier, Marianne: ILVO Seeds project manager with experience in breeding*
- *Marchand, Fleur: Researcher in interdisciplinary systems thinking–Scientific Director*
- *Martinez Ruiz, Valentina: Agricultural economist and researcher in rural development*
- *Maselyne, Jarissa: Researcher in precision livestock farming*
- *Muylle, Hilde: Researcher in new crop development and supply chain innovation–Director*
- *Nuyttens, David: Precision agriculture researcher with expertise in innovative crop protection techniques*
- *Pannecoucque, Joke: Researcher in new crop development*
- *Polet, Hans: Researcher in sustainable fishing techniques–Scientific Director*
- *Rasschaert, Geertrui: Food microbiologist with expertise in pathogens and molecular techniques*
- *Reubens, Bert: Researcher in nature-inclusive farming techniques with expertise in agroforestry*
- *Robbens, Johan: Marine biotechnologist and algae expert*
- *Rogge, Elke: Researcher in rural development and spatial policy–Scientific Director*
- *Roldán-Ruiz, Isabel: Researcher in molecular breeding, genomics and high-throughput plant phenotyping–Scientific Director*
- *Ruysschaert, Greet: Soil and climate expert; researcher in sustainable tillage techniques*
- *Sonck, Bart: Researcher in precision livestock farming– Head of Department*

- *Tuyttens, Frank: Animal welfare researcher*
- *Van Beek, Jonathan: Researcher in precision agriculture, image analysis and drone technology*
- *Van Droogenbroeck, Bart: Food technologist and researcher on the valorization of food by-products*
- *Van Huylenbroeck, Johan: Agricultural engineer with a passion for breeding and ornamental horticulture–Scientific Director*
- *Van Laecke, Kristiaan: Agricultural engineer with expertise in plant breeding and crop protection–Head of Department*
- *Van Laere, Katrijn: Plant breeding researcher and expert in new genomic techniques*
- *Van linden, Veerle: Researcher in environmental sustainability and expert in life cycle analysis*
- *Van Pamel, Els: Researcher in chemical food safety and quality*
- *Van Renterghem, Hanna: Communications officer at the B2BE Facilitator and member of the ILVO communications team*
- *Van Royen, Geert: Researcher in protein diversification and food applications–Director*
- *Van Vaerenbergh, Johan: Expert in plant health and plant-pathogenic bacteria*
- *Van Weyenberg, Stephanie: Expert in digitalization and data sharing in the agrifood industry–Director–Coordinator of DjustConnect*
- *Vanden Nest, Thijs: Manager of ILVO's experimental farm and crop production expert*
- *Vandermaelen, Hans: Spatial researcher with expertise in public agricultural land and the relationship between urbanization, agriculture and food*
- *Vanderperren, Els: Researcher in sustainable fishing techniques–Director*
- *Vanempten, Elke: Researcher in productive landscapes and spatial planning–Director*
- *Vangeyte, Jürgen: Researcher in precision agriculture and data technology–Scientific Director*
- *Vanhoorne, Sam: AI developer for marine monitoring applications*
- *Verhoeve, Anna: Spatial researcher with expertise in urbanization and rural development*
- *Verstraete, Karen: Food technologist and communications officer at the Food Pilot, The ProteInn Club and NuHCaS*
- *Versyck, Jasmine: Agricultural engineer and business developer at the B2BE Facilitator*
- *Vlaemynck, Geertrui: Researcher with expertise in nutrition for patients and groups with special needs*
- *Vleugels, Tim: Plant breeding researcher*

- *Wauters, Erwin: Agricultural economist with expertise in risk management and research into new business models*
- *Willekens, Koen: Researcher in soil management and in organic and agroecological farming systems and compost specialist*
- *Willems, Glenn: Spatial researcher with a focus on landscape systems*

In cooperation with the Center for Agrarian History (CAG)

Bibliography

Articles and Reports

- *Agentschap Landbouw en Zeevisserij. (2024). Landbouwrapport 2024 (LARA): Vlaamse landbouw in cijfers. https://landbouwcijfers.vlaanderen.be/landbouwrapport-2024-lara*
- *Avansya. (2024). The environmental footprint of EverSweet®: A comparative life cycle assessment of commercial sweeteners in 2023. https://www.avansya.com/wp-content/uploads/2024/04/EverSweet-LCA-Summary_FINAL.pdf*
- *Battle, M., Bomkamp, C., Carter, M., Colley Clarke, J., Eastham, L., Fathman, L., Gertner D., Kirchner J., Harsini F., Leet-Otley T., & Swartz, E. (2024). State of the industry: Cultivated meat, seafood, and ingredients. The Good Food Institute. https://gfi.org/resource/cultivated-meat-seafood-and-ingredients-state-of-the-industry/*
- *Cian, R. E., Fajardo, M. A., Alaiz, M., Vioque, J., González, R. J., & Drago, S. R. (2014). Chemical composition, nutritional and antioxidant properties of the red edible seaweed* Porphyra columbina. *Food Chemistry, 165, 299–305.*
- *Delbrassinne, L., Verhaegen, B., Van Damme, I., & Van Hoorde, K. (2024). Voedseltoxi-infecties in België en Vlaanderen: Jaaroverzicht 2023. Sciensano. https://www.sciensano.be/sites/default/files/jaarverslag_vti_2023_nl_2024_vlaanderen_finaal.pdf*
- *Don, A., Seidel, F., Leifeld, J., Kätterer, T., Martin, M., Pellerin, S., Emde, D., Seitz, D., & Chenu, C. (2023). When does soil carbon contribute to climate change mitigation? Policy brief. EJP Soil. https://doi.org/10.5281/zenodo.13970665*
- *Facq, E., & De Waegemaeker, J. (2025). Carbon farming in Vlaanderen: Verkenning van een aanvullend verdienmodel (ILVO beleidsadvies 2025.04). Instituut voor Landbouw-, Visserij- en Voedingsonderzoek.*
- *FAO. 2025. The Third Report on the State of the World's Plant Genetic Resources for Food and Agriculture. FAO Commission on Genetic Resources for Food and Agriculture. https://doi.org/10.4060/cd4711en*
- *Foley, J. A., Ramankutty, N., Brauman, K. A., Cassidy, E. S., Gerber, J. S., Johnston, M., Mueller, N. D., O'Connell, C., Ray, D. K., West, P. C., Balzer, C., Bennett, E. M., Carpenter, S. R., Hill, J., Monfreda, C., Polasky, S., Rockström, J., Sheehan, J., Siebert, S., Tilman, D., & Zaks, D. P. M. (2011). Solutions for a cultivated planet. Nature, 478(7369), 337–342. https://www.nature.com/articles/nature10452*
- *Giller, K. E., Hijbeek, R., Andersson, J. A., & Sumberg, J. (2021). Regenerative agriculture: An agronomic perspective. Outlook on Agriculture, 50(1), 13–21. https://doi.org/10.1177/0030727021998063*

- *Gobin, A., & Van de Vijver, H. (2021). Spatio-temporal variability of dry and wet spells and their influence on crop yields. Agricultural and Forest Meteorology, 308–309, 108565.* https://doi.org/10.1016/j.agrformet.2021.108565
- *Grafton, R. Q., Williams, J., Perry, C. J., Molle, F., Ringler, C., Steduto, P., Udall, B., Wheeler, S. A., Wang, Y., Garrick, D., & Allen, R. G. (2018). The paradox of irrigation efficiency: Higher efficiency rarely reduces water consumption. Science, 361(6404), 748–750.* https://doi.org/10.1126/science.aat931
- *Living Lab Agro-ecologie en Biologische Landbouw. (2024). De kracht van samenwerking in eerlijke voedselketens. ILVO, BioForum, CCBT, Voedsel Anders & Steunpunt Korte Keten.* https://ilvo.vlaanderen.be/uploads/documents/De-kracht-van-samenwerking-in-eerlijke-voedselketens.pdf
- *Lowder, S. K., Sánchez, M. V., & Bertini, R. (2021). Which farms feed the world and has farmland become more concentrated? World Development, 142, 105455.* https://doi.org/10.1016/j.worlddev.2021.105455
- *Martinez Ruiz, V., Van Caenegem, E., Messely, L., & Marchand, F. (2024). De werkelijke prijs van voedsel (ILVO beleidsadvies 2024.04). Instituut voor Landbouw-, Visserij- en Voedingsonderzoek.*
- *Nakagaki, B. J., Sunde, M. L., & DeFoliart, G. R. (1987). Protein quality of the house cricket,* Acheta domesticus, *when fed to broiler chicks. Poultry Science, 66, 1367–1371.*
- *Prusinski, J. (2017). White lupin (Lupinus albus L.) – Nutritional and health values in human nutrition: A review. Czech Journal of Food Science, 35(2), 95–105.*
- *Reubens, B., D'Haene, K., D'Hose, T., & Ruysschaert, G. (2010). Bodemkwaliteit en landbouw: Een literatuurstudie. Activiteit 1 van het Interregproject BodemBreed. Instituut voor Landbouw-, Visserij- en Voedingsonderzoek (ILVO).*
- *Rubens, K., Wynants, E., Moyersoen, I., Guffens, L., & Michiels, K. (2024). EI-MEET: Monitoring eiwitinname en -aankopen in Vlaanderen 2023. Departement Omgeving, Vlaamse overheid.* https://publicaties.vlaanderen.be/view-file/73755
- *Safi, C., Zebib, B., Merah, O., Pontalier, P.-Y., & Vaca-Garcia, C. (2014). Morphology, composition, production, processing and applications of* Chlorella vulgaris: *A review. Renewable and Sustainable Energy Reviews, 35, 265–278.*
- *Souci, S. W., Fachmann, W., & Kraut, H. (2008). Food composition and nutrition tables (7th rev. ed.). CRC Press.*
- *Tits, M., Elsen, A., Deckers, S., Bries, J., & Vandendriessche, H. (2020). Bodemvruchtbaarheid van de akkerbouw- en weilandpercelen in België en Noordelijk Frankrijk (2016–2019). Bodemkundige Dienst van België.*
- *Tubb, C., & Seba, T. (2019). Rethinking food and agriculture 2020–2030: The second domestication of plants and animals, the disruption of the cow, and the collapse of industrial livestock farming. RethinkX.* https://www.rethinkx.com/publications/rethinkingfoodandagriculture2019.en

- *Tuyttens, F. A. M., Molento, C. F. M., & Benaissa, S. (2022). Twelve threats of precision livestock farming (PLF) for animal welfare. Frontiers in Veterinary Science, 9, 889623. https://doi.org/10.3389/fvets.2022.889623*
- *U.S. Dairy Export Council. (2005). Reference manual for U.S. milk powders. U.S. Dairy Export Council.*
- *Vandermaelen, H. (2023). Urbanising the agroecological reproduction of soil fertility (Master's thesis, Ghent University, Faculty of Engineering and Architecture). http://hdl.handle.net/1854/LU-01GRV4PF6Q6QPNBE3FDWZRVEW7*
- *Vandermaelen, H., Willems, G., & Vanempten, E. (2024). Publieke grond in Vlaanderen en Brussel: Overzicht, evolutie en landbouwpotentieel. Instituut voor Landbouw-, Visserij- en Voedingsonderzoek.*
- *Vandermaelen, H., Willems, G., Verhoeve, A., Vanempten, E., & De Waegemaeker, J. (2024). Een actief grond- en pandenbeleid voor grondgebonden landbouw (ILVO beleidsadvies 2024.07.01). Instituut voor Landbouw-, Visserij- en Voedingsonderzoek.*
- *Verhoeve, A. (2015). Revealing the use of farms and farmland by non-agricultural economic activities: The case of Flanders (Doctoral dissertation, KU Leuven, Faculty of Bioscience Engineering). https://doi.org/10.13140/RG.2.1.2696.3047*
- *Vlaamse Landmaatschappij. (2024). Mestrapport 2024. Vlaamse overheid. https://publicaties.vlaanderen.be/view-file/71650*
- *Vlaams Instituut voor Biotechnologie. (2019). Precisieveredeling in planten via CRISPR-Cas (VIB Fact Series). https://vib.be/sites/vib.sites.vib.be/files/2023-10/vib_CRISPR-Cas_NL_2019_0118_LR.pdf*
- *Wieme, J., Leroux, S., Cool, S. R., Van Beek, J., Pieters, J. G., & Maes, W. H. (2024). Ultra-high-resolution UAV-imaging and supervised deep learning for accurate detection of* Alternaria solani *in potato fields. Frontiers in Plant Science, 15, 1206998. https://doi.org/10.3389/fpls.2024.1206998*
- *Witte, B., Obloj, P., Koktenturk, S., Morach, B., Brigl, M., Rogg, J., Schulze, U., Walker, D., Von Koeller, E., Dehnert, N., & Grosse-Holz, F. (2021). Food for thought: The protein transformation. Boston Consulting Group. https://www.bcg.com/publications/2021/the-benefits-of-plant-based-meats*
- *Wullaert, A., Byttebier, K., & Smets, V. (2024). Economische impact van de alternatieve eiwitsector in Vlaanderen en Brussel. Departement Omgeving, Vlaamse overheid. https://publicaties.vlaanderen.be/view-file/69017*
- *Zielinska, E., Barniak, B., Karas, M., Rybczynska, K., & Jakubczyk, A. (2015). Selected species of edible insects as sources of nutrient composition. Food Research International, 77, 460–466.*

Books

- *Dries, L. (Ed.). (2019). EU bioeconomy economics and policies: Volume I and II. Palgrave Macmillan.*
- *Eise, J., & Foster, K. (Eds.). (2018). How to feed the world. Island Press.*
- *Josling, T. E., & Tangermann, S. (2015). Transatlantic food and agricultural trade policy: 50 years of conflict and convergence. Edward Elgar Publishing.*
- *Koning, N. (2017). Food security, agricultural policies and economic growth. Routledge.*
- *Mann, C. C. (2019). The wizard and the prophet. Picador.*
- *Matthijs, E., & Relaes, J. (2012). Landbouw en voedsel verrassend actueel. Acco.*
- *Meester, G. (Ed.). (2013). EU-beleid voor landbouw, voedsel en groen. Wageningen Academic Publishers.*
- *Parmentier, B. (2007). Nourrir l'humanité: Les grands problèmes de l'agriculture mondiale au XXIe siècle. La Découverte.*
- *Ritchie, H. (2024). Niet het einde van de wereld. Uitgeverij Balans.*
- *Steel, C. (2008). Hungry City: How Food Shapes Our Lives. Chatto & Windus.*
- *Swinnen, J. (Ed.). (2015). The political economy of the 2014–2020 Common Agricultural Policy. Routledge.*
- *Veerman, D., & Gravensteijn, E. (Eds.). (2025). Daarom zijn boeren boos. Prometheus.*
- *Wetenschappelijke Raad voor het Regeringsbeleid. (2024). Naar een voedselbeleid. Amsterdam University Press.*

Websites and online databases

- *Agentschap Landbouw en Zeevisserij. (2025). Landbouwcijfers. Vlaamse overheid. https://www.vlaanderen.be/landbouwcijfers*
- *Algemene Directie Statistiek – Statistics Belgium. (2023, september). Nieuwe resultaten Belgisch huishoudbudgetonderzoek. Federale Overheidsdienst Economie. https://statbel.fgov.be/nl/themas/huishoudens/huishoudbudget*
- *Algemene Directie Statistiek – Statistics Belgium. (2023). Landbouwcijfers. Federale Overheidsdienst Economie. https://statbel.fgov.be/nl/themas/landbouw-visserij/*
- *Cacaolab. (2025). Cacao alternatives. https://www.cacaolab.be/cocoa-alternatives*
- *CaixaBank Research. (2023, October). Production costs and drought are affecting Spain's agrifood sector. Sectoral analysis – Agrifood. https://www.caixabankresearch.com/en/sectoral-analysis/agrifood/production-costs-and-drought-are-affecting-spains-agrifood-sector*
- *Departement Omgeving. (2023). Ruimtebeslag. Vlaamse overheid. https://indicatoren.omgeving.vlaanderen.be/indicatoren/ruimtebeslag*

- De Winter, H. (2015). *175 jaar bemestingsleer en 70 jaar Bodemkundige Dienst van België: Over de verspreiding van bemestingskennis.* Centrum Agrarische Geschiedenis. https://cagnet.be/page/bemestingsleer
- Directoraat-generaal Klimaat. (n.d.). *Gevolgen van de klimaatverandering.* Europese Commissie. https://climate.ec.europa.eu/climate-change/consequences-climate-change_nl
- Garré, S. (2024, 4 March). *Gezuiverd afvalwater voor irrigatie: Beter voorkomen dan genezen.* EOS Blogs. https://www.eoswetenschap.eu/natuur-milieu/gezuiverd-afvalwater-voor-irrigatie-beter-voorkomen-dan-genezen
- Huysmans, M. (2020, 15 April). *Pompen landbouw en industrie in Vlaanderen veel meer grondwater op dan er aangevuld wordt? Mythes over grondwater.* https://grondwatermythes.blogspot.com/2020/04/pompen-landbouw-en-industrie-in.html
- Instituut voor Landbouw-, Visserij- en Voedingsonderzoek. (z.d.). *ILVO – Website.* https://www.ilvo.vlaanderen.be
- IPM Trips. (2020–2024). *Beheersing: Irrigatie.* ILVO, Viaverda, Inagro & Proefstation voor de Groenteteelt. https://www.ipmtrips.be/nl/beheersing/irrigatie
- Katholieke Universiteit Leuven. (2025, 2 april). *KU Leuven ontwikkelt technologie die bananen genetisch weerbaarder kan maken.* https://nieuws.kuleuven.be/nl/2025/ku-leuven-ontwikkelt-technologie-die-bananen-genetisch-weerbaarder-kan-maken
- Mondelēz International. (2025). *Het Harmony Programma.* https://www.harmony.info/nl-be/
- Segers, Y., Woestenborghs, B., & Bekaert, J. (2004). *In de greep van de vooruitgang (1880–1950).* Centrum Agrarische Geschiedenis. https://cagnet.be/page/landbouw-1880-1950-vooruitgang
- Statistiek Vlaanderen. (2023). *Zorg en ondersteuning voor ouderen.* Vlaamse overheid. https://www.vlaanderen.be/statistiek-vlaanderen/zorg/zorg-en-ondersteuning-voor-ouderen
- Testaankoop. (2025, 5 March). *Nu de prijsonderhandelingen aanslepen: "Koffie werd in 3 jaar tijd al 29% duurder," zegt Testaankoop.* https://www.test-aankoop.be/familie-prive/supermarkten/pers/inflatie-februari-2025
- Vlaams Centrum voor Agro- & Visserijmarketing. (n.d.). *Kennisbank.* https://www.vlaanderen.be/vlam/kennisbank
- Vlaams Infocentrum voor Land- en Tuinbouw. (n.d.). *VILT – Website.* https://www.vilt.be
- Vlaamse Milieumaatschappij. (2025). *Feiten & cijfers: Water.* Vlaamse overheid. https://vmm.vlaanderen.be/feiten-cijfers/water

Presentations

- *Atzori, G., & Garré, S. (2024, April 16). Crops for saline farming. [session presentation]. SALAD-SUSTAIN Conference, Saline Agriculture: State of the art and the prospects of impact investment, Brussels. https://ilvo.vlaanderen.be/uploads/documents/SALAD/2_Crops-for-saline-farming.pdf*
- *Martellozzo, F., Randelli, F., Ferrone, L., Vaglie, M. D., Falaguasta, C., & others. (2024, April 16). Upscaling geographical impacts of extreme sea level rise and salinization. [session presentation]. SALAD-SUSTAIN Conference, Saline Agriculture: State of the art and the prospects of impact investment, Brussels. https://ilvo.vlaanderen.be/uploads/documents/SALAD/5_Climate-scenarios-and-salinity-maps.pdf*
- *Van Royen, G. (2025, 8 mei). Challenges of alternative proteins and other ingredients for innovative, nutritional and safe food. In Nourishing a growing world: Food transition for a healthy planet and healthy lives [session presentation]. Knowledge for Growth, Flanders Meeting & Convention Center, Antwerpen.*
- *Willekens, K., & Delanote, L. (2025, 10 februari). Een vruchtbare bodem in vier dimensies ontleed [lecture]. Winterbijeenkomst over bemesting en bodemvruchtbaarheid in bio, Pamel.*

Journals

- *Bambridge-Sutton, A. (2025, 30 mei). Is regenerative agriculture driven by consumer demand? Food Navigator. https://www.foodnavigator.com/Article/2025/05/30/nestle-and-danone-on-demand-for-regenerative-agriculture/*
- *Demeulemeester, S. (2025, 7 juni). Burgeronderzoek Watermonsters geeft zwemmers hoop: Grote waterlopen schoner dan verwacht. De Standaard. https://www.standaard.be/binnenland/burgeronderzoek-watermonsters-geeft-zwemmers-hoop-grote-waterlopen-schoner-dan-verwacht/69854662.html*
- *De Cleene, D. (2024, 5 augustus). Kaas die de wereld verandert: Is deze 'stalen koe' van Gents bedrijf de sleutel tot duurzaamheid? De Morgen. https://www.demorgen.be/beter-leven/kaas-die-de-wereld-verandert-is-deze-stalen-koe-van-gents-bedrijf-de-sleutel-tot-duurzaamheid~b130e82e/*
- *Fresco, L. O. (2025, 16 juni). De wereld ten onder? Nee. Alles komt goed? Ook niet. NRC. https://www.nrc.nl/nieuws/2025/06/16/de-wereld-ten-onder-nee-alles-komt-goed-ook-niet-a4897100*
- *Tanis, M. (2012, mei). Dierlijke restverwerking: Niets blijft ongebruikt. Veeteelt. https://edepot.wur.nl/211109*
- *Verwimp, B. (2022, 28 december). En de boer, hij ploeterde voort. De Standaard.*
- *Veldverkenners. (2015). Terug in de tijd met Veldverkenners. Vlaams Infocentrum voor Land- en Tuinbouw.*
- *Verstraete, K., Vlaeminck, S., Van Royen, G., Ameloot, N., & Waegeman, H. (2023). Fermentatiegebaseerde eiwitten ontwikkelen, produceren en verwerken tot food, feed en fijnchemicaliën. Food Science & Law, 1, 2–15. Die Keure.*